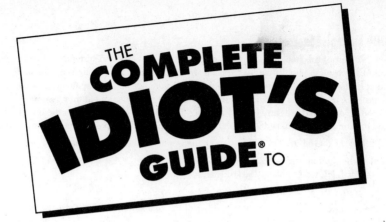

Geometry

by Denise Szecsei, Ph.D.

ALPHA

A member of Penguin Group (USA) Inc.

D0753839

International Standard Book Number: 1-59257-183-2
Library of Congress Catalog Card Number: 2004100483

06 05 04   8 7 6 5 4 3 2 1

Interpretation of the printing code: The rightmost number of the first series of numbers is the year of the book's printing; the rightmost number of the second series of numbers is the number of the book's printing. For example, a printing code of 04-1 shows that the first printing occurred in 2004.

Printed in the United States of America

Note: This publication contains the opinions and ideas of its author. It is intended to provide helpful and informative material on the subject matter covered. It is sold with the understanding that the author and publisher are not engaged in rendering professional services in the book. If the reader requires personal assistance or advice, a competent professional should be consulted.

The author and publisher specifically disclaim any responsibility for any liability, loss, or risk, personal or otherwise, which is incurred as a consequence, directly or indirectly, of the use and application of any of the contents of this book.

Most Alpha books are available at special quantity discounts for bulk purchases for sales promotions, premiums, fund-raising, or educational use. Special books, or book excerpts, can also be created to fit specific needs.

For details, write: Special Markets, Alpha Books, 375 Hudson Street, New York, NY 10014.

Publisher: *Marie Butler-Knight*
Product Manager: *Phil Kitchel*
Senior Managing Editor: *Jennifer Chisholm*
Senior Acquisitions Editor: *Mike Sanders*
Development Editor: *Michael Koch*
Senior Production Editor: *Billy Fields*
Copy Editor: *Molly Schaller*
Illustrator: *Chris Eliopoulos*
Cover/Book Designer: *Trina Wurst*
Indexer: *Heather McNeil*
Layout/Proofreading: *Becky Harmon, Mary Hunt*

Contents at a Glance

Part 1:   **The Foundation**   **1**

   1   What Is Geometry, Anyway?   3
Everyone has heard of Euclid, but for most people, geometry is Greek to them.

   2   Let's Do Algebra   13
Brush up your algebra skills, as you will need them on this geometric journey.

   3   Building Blocks   25
Learn the lingo: the first defined and undefined terms, as well as a postulate or two to get things rolling.

   4   There's Always an Angle   37
Learning angle classifications and sorting out the various relationships.

   5   Lines and the Angles They Form   49
Linear interactions are fairly restrictive, but magical things can still happen.

   6   A Polygon Is a Many-Sided Thing!   59
A polygon by any other name would smell as sweet.

Part 2:   **Introducing Proofs**   **71**

   7   Logic: Rules for Arguing   73
"It is too!" might be enough to convince your younger brother, but it doesn't carry much weight here.

   8   Taking the Burden out of Proofs   85
The proof is in the pudding, and this is where you'll get your just desserts!

   9   Proving Segment and Angle Relationships   97
Cutting up segments and angles with razor-sharp postulates.

   10   Proving Relationships Between Lines   109
Sometimes when two lines meet, everything is just right.

Part 3:   **Piecing Together Triangles and Quadrilaterals**   **123**

   11   Two's Company, Three's a Triangle   125
Mathematical triangles: No one is jealous and no one gets hurt.

12 Congruent Triangles   139
*There are more acronyms here than there were during FDR's
New Deal!*

13 Similar Triangles   155
Keep it in proportion. Remember, it's all about the angles.

14 Opening Doors with Similar Triangles   167
Your first proof of the Pythagorean Theorem.

15 Putting Quadrilaterals in the Forefront   179
Those famous four-sided figures take center stage.

16 Proofs About Quadrilaterals   195
Play an exciting new game show and win fabulous prizes.

Part 4:   Going Around in Circles   **207**

17 Anatomy of a Circle   209
Dissecting a circle is as easy as pi.

18 Segments and Angles   225
Strumming the chords together to make beautiful music.

19 Circular Arguments   237
These proofs about circles will make your head spin.

20 The Unit Circle and Trigonometry   251
*Going around and around: from triangles to circles back to
triangles again.*

Part 5:   Where Can We Go from Here?   **267**

21 The Next Dimension: Surfaces and Solids   269
*That's the signpost up ahead. Your next stop, the Twilight
Zone.*

22 Under Construction   281
You won't need a hard hat on this construction site.

23 When Geometry and Algebra Intersect   293
Geometry with a French accent.

24 Whose Geometry Is It Anyway?   305
Go beyond Euclid's wildest dreams.

25 Transformations   317
*The captain has turned off the "Fasten Seat Belt Sign" and it
is now safe to move around the plane.*

Appendixes

A Answer Key   331

B Postulates and Theorems   351

C Formulas   357

D Glossary   359

   Index   367

Contents

Part 1: The Foundation   **1**

1 What Is Geometry, Anyway?   **3**

What's the Point? 4
  How Does Geometry Measure Up? 4
  Getting Into Shape 5
  Let's Be Reasonable: Learning How to Write Proofs 6
Who's Idea Was It? 6
  The Tale of Thales 6
  And the Winner Is ... Euclid 7
  David Hilbert: A Goliath Mathematician 9
Can I Really Learn This? 10

2 Let's Do Algebra   **13**

Long Lost Relations 14
Properties of Equality 15
  Reflexive, Symmetric, and Transitive Properties 16
  Algebraic Properties of Equality 17
  The Square Root Property of Equality 19
Properties of Inequality 20
  Is It an Equivalence Relation? 21
  Algebraic Properties of Inequality 21
  An Additional Additive Property 22

3 Building Blocks   **25**

Coming to Terms 26
Points and Space 26
Lines 28
Line Segments 29
  Segment Length 30
  A New Relation 31
  Segment Addition 32
  Between 32
Rays 33
Planes 34

4   There's Always an Angle   37

What's in a Name? 38

Are You My Type? The Basic Angle Classifications 40

Angle Addition 41

How Do Angles Relate? Classifying Pairs of Angles 42

    Adjacent Angles 42

    Congruent Angles 43

    Complementary and Supplementary Angles 44

Algebraic Games We Can Play 44

A First Look at Proving Angle Congruence 46

5   Lines and the Angles They Form   49

Linear Interactions 50

Intersecting Lines and Vertical Angles 50

Perpendicular Lines 52

Parallel Lines: Euclid's 5th 55

Transversals and the New Angle Pairs 57

6   A Polygon Is a Many-Sided Thing!   59

Coming to Terms with the Terminology 60

Naming Conventions and Classifications 63

The Interior Angles 64

Regular Polygons 67

Part 2:   Introducing Proofs   71

7   Logic: Rules for Arguing   73

Inductive Reasoning 74

Deductive Reasoning: Elementary, My Dear Watson! 75

Logical Constructions and Truth Tables 76

    Negation 76

    Conjunction 77

    Disjunction 78

    Implication (or Conditional) 79

    Logical Equivalence and Tautology 81

8   Taking the Burden out of Proofs   85

The Law of Detachment 86

The Importance of Being Direct 87

Proof by Contradiction: The Advantage of Being Indirect 89

The Given Information: Use It or Lose It! 90
A Solid Foundation: Definitions, Postulates, and Theorems 90
What Should You Bring to a Formal Proof? 92

9 Proving Segment and Angle Relationships   97

Exploring Midpoints 98
How Many Midpoints Are There? 99
Proving Angles Are Congruent 101
Using and Proving Angle Complements 102
Using and Proving Angle Supplements 105

10 Proving Relationships Between Lines   109

Proofs Involving Perpendicular Lines 110
Let's Get Parallel 112
Proofs About Alternate Angles 113
Parallel Lines and Supplementary Angles 115
Using Parallelism to Prove Perpendicularity 116
Proving Lines Are Parallel 117

Part 3:   Piecing Together Triangles and Quadrilaterals   123

11 Two's Company, Three's a Triangle   125

A Formal Introduction 126
  Classifying Triangles by Their Angles or Their Sides *126*
  Sums of Interior Angles Are Cooking at 180° *128*
  Exterior Angle Relationships *130*
Size Matters, So Let's Measure 132
  Perimeter *132*
  Area *132*
  Ladies and Gentlemen: The Pythagorean Theorem *134*
  The Triangle Inequality *135*

12 Congruent Triangles   139

CPOCTAC 140
The Big Five 140
  The SSS Postulate *141*
The SAS Postulate 142
The ASA Postulate 144
The AAS Theorem 145
  The HL Theorem for Right Triangles *146*
Proving Segments and Angles Are Congruent 149
Proving Lines Are Parallel 150

13 Similar Triangles   **155**

Ratio, Proportion, and Geometric Means 156

Properties of Similar Triangles 159

The Big Three 161

   The AAA Similarity Postulate *161*

   The SAS and SSS Similarity Theorems *164*

14 Opening Doors with Similar Triangles   **167**

The Pythagorean Theorem 168

Parallel Segments and Segment Proportions 170

Three Famous Triangles 173

   60-60-60 Triangle *173*

   30-60-90 Triangle *174*

   45-45-90 Triangle *176*

15 Putting Quadrilaterals in the Forefront   **179**

Properties of All Quadrilaterals 180

Properties of Trapezoids 180

Let's All Fly a Kite! 182

Properties of Parallelograms 184

The Most Popular Parallelograms 186

   Rectangles *186*

   Rhombuses *187*

   Squares *189*

Calculating Areas 189

   Area of Trapezoids *189*

   Area of Parallelograms *190*

   The Pythagorean Theorem (again) *191*

16 Proofs About Quadrilaterals   **195**

When Is a Quadrilateral a Parallelogram? 196

   Opposite Sides Congruent and Parallel *196*

   Two Pairs of Congruent Sides *197*

   Two Pairs of Congruent Angles *199*

   Bisecting Diagonals *200*

When Is a Parallelogram a Rectangle? 201

When Is a Parallelogram a Rhombus? 202

When Is a Parallelogram a Square? 204

Part 4:   Going Around in Circles   207

17   Anatomy of a Circle   209

Basic Terms 210
Arcs 211
Circumference and Area: Pi Anyone? 215
Tangents 217
From One Theorem Comes Many 218

18   Segments and Angles   225

Angles and Chords 226
Arcs and Chords 227
Radii and Chords 229
Putting the Pieces Together 233

19   Circular Arguments   237

Angles and Arcs 238
Similarity 241
Parallel Chords and Arcs 245
Putting Your Problems Behind You 247

20   The Unit Circle and Trigonometry   251

The Tangent Ratio 252
The Sine Ratio 255
The Cosine Ratio 257
And the Rest 259
How Does This Relate to the Unit Circle? 261

Part 5:   Where Can We Go from Here?   267

21   The Next Dimension: Surfaces and Solids   269

Prisms 270
Pyramids 273
Cylinders and Cones 274
Polyhedra 276
Spheres 277
Platonic Solids 278

22 Under Construction **281**

Tools of the Trade 282
    Straight Edge 282
    Compass 283
    Protractors 283
Bisection 284
    Bisecting Segments 284
    Bisecting Angles 285
Constructing Lines 287
    Perpendicular Lines 287
    Parallel Lines 289
Circular Constructions: Off on Yet Another Tangent 289
Constructing Quadrilaterals 290
    Kites 290
    Parallelograms 291
    Rectangles 291
    Rhombuses 291
    Squares 292

23 When Geometry and Algebra Intersect **293**

The Cartesian Coordinate System 294
Finding Horizontal and Vertical Distances 295
The Pythagorean Theorem Goes the Distance 296
The Midpoint Formula 297
Finding Equations of Lines 298
    The Slope of a Line: Not Another Tangent! 298
    The Point-Slope Formula 300
    The Intercepts 300
The Secret Lives of Parallel and Perpendicular Lines 301
Graphing Lines: A Picture Is Worth a Thousand Words 302

24 Whose Geometry Is It Anyway? **305**

Non-Euclidean Geometry 306
Hyperbolic Geometry: Saddle Up! 307
Spherical Geometry 309
Taxi-Cab Geometry 310
Max Geometry 312
How Many Shapes Can a Circle Have? 313

25 Transformations   **317**

Isometries 318
 Translations *319*
 Reflections *320*
 Rotations *322*
 Glide Reflections *324*
Dilations 325
Symmetry 326

Appendixes

A Answer Key   **331**

B Postulates and Theorems   **351**

C Formulas   **357**

D Glossary   **359**

Index   **367**

Foreword

The first phase of learning anything new is the idiot phase. Remember learning to play tennis? To play the piano? To read? It's humbling to go through those beginning idiot steps. The goal of learning something new has to be strong to pull us along.

Why should we learn geometry? We live in a three-dimensional world that is filled with the patterns and structure of geometry. We see geometry in the shapes of our buildings, furnishings, art, sculpture, landscape design and countless other objects. We see the beauty of geometry not only in man-made products, but in the structure of crystals, flowers, snowflakes, and numerous aspects of nature. We measure our world using geometric constructs. Global positioning systems help us to determine exact locations in airplanes, cars and even golf carts. Our precise measurements enabled us to accomplish great achievements like exploring Mars with sophisticated robotic machines. Geometry is a very useful subject.

In nearly every subject we study, we are presented with a large number of facts. This is not true of geometry (and there is good reason for it). In geometry we start at the very beginning with the undefined terms of point, line, and plane and we build logical structure using a very few statements which are accepted as true without proof. These statements called postulates and axioms, along with some defined terms, help us to explore and discover geometric relationships. We do this by constructing and proving theorems about those relationships. Thus, the tools of geometry are revealed one a time by proof. The process of proof may be the most useful tool of all.

In Euclidean Geometry, which is what this book is about, there are no controversies. By using this system of proof, we can be certain of our conclusions. A huge side benefit of this approach to learning geometry, is that the system can teach us to reason logically. Logical thinking is applicable to every learning discipline.

Learning to "do" geometric proof requires a new way of thinking. Like anything else worthwhile, it will take some time to understand how the system works. You didn't learn to walk or read or swim overnight. You won't feel comfortable with proofs right away. You won't remember all of the postulates and definitions. Hang in there. It will come. Draw pictures of your problem. Ask yourself, "What do I know about this figure?" Write it down. Eventually you will start thinking "I wonder if …" and " Why?" You will have graduated past the idiot stage.

This book covers all of the topics of any standard geometry text. You will be pleased to note, however, that it is a bit different than those textbooks. It is nonthreatening.

The author does a masterful job of walking you through the material in a user-friendly way. Her great sense of humor and comparisons to real-world situations make this book a pleasure. This book provides the basic concepts that will enable you to use and explore the geometry you encounter in your life. Hopefully, you will also discover that geometry is truly beautiful.

—Dale Seymour
Geometry author, publisher, and instructor

Introduction

I'll admit this right away: I like math. I like problem solving, working with numbers, and thinking logically. I know that it is one of the many quirky things about me; there are many people who do not share my enthusiasm for the subject. I've met enough people that struggle with math to realize that most people would rather face the challenges on *Fear Factor* than take a math test. This book was written with those folks in mind.

At this point in your mathematical career you might have just gotten comfortable with algebra. I use the word "comfortable" loosely; you won't hyperventilate if you have to solve for x. Algebra is familiar, and you know what to expect. Now that you are about to study geometry you might feel as if you are being forcefully relocated to a foreign, hostile environment.

When you listen to what geometry survivors have to say, you'll get the impression that geometry is foreign, confusing, and treacherous. This is very odd, because most people have learned basic ideas in geometry by watching clever programs on public television stations. Most preschoolers get excited about geometry. They love to point out the triangles, circles, and squares around the house and in the yard. They love to eat triangular sandwiches and square crackers, and drink out of cylindrical sippie-cups. Food and geometry go hand-in-hand, and you can't go wrong with that!

But somewhere along the way geometry turned into a foreign language. Maybe it's because Greek letters like α and β enter into the mix instead of the good old-fashioned last letters of the alphabet that are used in algebra. Don't let the Greek alphabet fool you. The Greeks write their words one letter at a time just like we do. In any case, you'll be relieved to know that in this book I will stick with the English alphabet and those friendly last letters of our alphabet!

Another intimidating thing about geometry is writing proofs. Sometimes it's hard enough just to solve a problem successfully. To have to explain and defend your work at every step is like jumping out of the frying pan and into the fire. But you shouldn't worry about this either. I'll walk you through the formalities of writing a proof, and I'll give you plenty of opportunities to go solo and show your skills in the exercises. Even when you are going it alone, I won't be far behind. The solutions to all of your problems (or at least your mathematical problems) can be found in Appendix A.

How This Book Is Organized

The material is presented in five parts.

Part 1, "The Foundation," introduces a bit of the history behind the subject. As a confidence booster, you'll have an opportunity to show off some of your algebra skills. This part also introduces you to some of the terminology, and you'll meet your first postulates (the ideas in geometry that we take on faith). Don't be afraid to use them throughout the book. That's what they live for. Then you'll learn the "ins and outs" of angles, lines, and polygons.

Part 2, "Introducing Proofs," explains how to formalize arguments. You will learn about different forms of reasoning and how to write a formal proof. You will practice your proof-writing skills on unarmed angles, segments, and lines.

Part 3, "Piecing Together Triangles and Quadrilaterals," examines a variety of polygons up close and personal. You will explore their properties and learn how to distinguish between true copies and counterfeit ones. Included with the purchase price of the book is a ticket to be a contestant in the new game show "Name That Quadrilateral."

Part 4, "Going Around in Circles," starts by dissecting circles and learning their parts. You will get a taste of pi, and then go on a triangle treasure hunt. You'll finish this part by drawing some connections between right triangles and circles.

Part 5, "Where Can We Go from Here?" takes you into the third dimension. You learn about some of the things that Homer Simpson briefly saw when he left his safe two-dimensional world of Springfield. You will also explore how to construct the geometric objects that I talk about in the earlier chapters. You'll have a chance to revisit the familiar world of algebra when you embed geometry in the Cartesian Coordinate System. You will leave Euclid's geometrical world and enter the strange world of non-Euclidean geometry. If you are not too exhausted, you'll move all of your geometric objects from one part of the plane to another. You won't have the resources to rent a truck or hire movers, so you'll have to do it all yourself. Fortunately these geometric objects weigh very little, so you shouldn't strain your muscles in the process. Remember to lift with your legs, not with your back.

Finally, the appendixes provide the answers to the mathematical problems in this book, as well as an at-a-glance summary of the postulates, theorems, and formulas provided.

Things to Help You Out Along the Way

When I talk about mathematics, there's always a tangent line I am tempted to go off on. Thanks to modern technology, I am not limited to discussing one thing at a time. I am able to put a parenthetical comment where it belongs, when it belongs. You'll see the following boxes throughout the book.

Tangled Knot

This box alerts you to the perils and dangers involved in jumping to conclusions, misusing notation, or being underinsured.

Solid Facts

This box provides definitions. Definitions are one of the key building blocks in geometry. It's important to use the right terms when you explain every step that is involved in your proofs.

Eureka!

This box contains helpful hints, useful observations, and the occasional "ah-ha!" moments.

Tangent Line

This box provides interesting trivia and strange but true facts. Although the subject matter is definitely interesting, these snippets are there to spice it up a bit.

Acknowledgments

There are many people who have helped my efforts to put these thoughts to paper, and although I can't list every one of them, I would like to name a few. First of all, I'd like to thank Erich Friedman for passing along that first e-mail message, and also for letting me use his compass and Platonic solids for a photo shoot. Margie Hale shared her experience regarding what to expect at the various stages of the writing process. Will Miles, Joe Lopez, Colin Branch and David Woods were instrumental in helping me figure out the problems with my figures. And the entire Mathematics and Computer Science Department at Stetson University encouraged me as I dove into this project.

I'd also like to thank Jessica Faust for making the connections that made this book happen, Mike Sanders for the initial okay, and Michael Koch for turning what I submitted into what you see before you.

On a more personal note, many of my family and friends chipped in to help me when I needed it.

Ross Szecsei was eager to look for errors, willing to help me draw the figures, and happy to take on extra chores as my deadlines drew near.

Ken Fishe came out of retirement to stop me from overusing the comma. Dan Oberlin helped me grow as a mathematician and as a teacher, and for that I will always be grateful.

Finally, I'd like to thank Kendelyn Michaels for helping me throughout this process. You'll never know how many corny jokes you were spared from reading because of her intervention! She was always willing to lend an ear and let me know when to leave it alone and when to rewrite.

Trademarks

All terms mentioned in this book that are known to be or are suspected of being trademarks or service marks have been appropriately capitalized. Alpha Books and Penguin Group (USA) Inc. cannot attest to the accuracy of this information. Use of a term in this book should not be regarded as affecting the validity of any trademark or service mark.

Part 1

The Foundation

The Greeks and geometry go hand and hand. They were the first to change the focus of learning from *how* to *why*. And that's what geometry is all about: learning why things are true. This part provides a brief look at the history of geometry.

Before you dive into proving all that you can prove, you will need to dust off your algebra skills. If you thought that studying geometry meant that you were done with algebra, think again. Algebra has the ability to work its way into places where you least expect it, and it will probably be years before you can finally put your algebra skills to rest.

After you learn the history of geometry and your algebra skills are where they need to be, you will be formally introduced to lines and angles, the building blocks of geometry.

What Is Geometry, Anyway?

In This Chapter

+ Why geometry is useful

+ The historic origins of geometry

+ Recent events in the development of geometry

+ The challenge of studying geometry

The Egyptians and the Babylonians were famous for their algebraic skills, and the mathematics they developed enabled them to develop clear instructions on *how* to do things. They were able to answer questions like: How can I add two numbers together? How can I find the volume of a pyramid? How can I learn more mathematics? Many of the Greek mathematicians traveled to Egypt to learn these mathematical skills. But after the Greeks learned *how* to do things, they became more interested in *why* they could do them. They asked themselves questions like: Why can't I draw a four-sided triangle? Why is the area of a rectangle the product of the length and the width? Why should I want to learn more mathematics?

Because this is the beginning of a book on geometry, it might be helpful if I talk a bit about what geometry is. The word *geometry* comes from the Greek words *geo* and *metron* which, when put together, mean "earth measure." This gives the impression that geometry involves measuring earthly things. So the next time you are asked "How on earth can we measure ..." you'll have a suggestion to make: Use geometry! Over time, the usage of the word geometry has evolved to describe an area of mathematics that studies the properties, measurement, and relationships of points, lines, angles, surfaces, and solids. This is a more accurate description of geometry, but as is typical, what one gains in accuracy, one loses in understandability!

Geometry is often found to be confusing and abstract. Writing it in English doesn't seem to help. Being introduced to it by singing purple dinosaurs and big yellow birds might have helped with the concrete ideas like shape and size (and improved geometry's image among preschoolers), but it has not helped develop an understanding of the abstract aspects of geometry. Let me assure you: The ideas in geometry are not too advanced for anyone to learn. After all, most of these ideas were well-understood thousands of years ago. Look at how much our civilization has advanced since then. If we can learn how to program a VCR, then surely we can tackle the material covered in geometry.

What's the Point?

Geometry has become entrenched in the standard preschool curriculum: You could fill a small library with the number of "board books" written about shape and size. You've probably been talking the geometry talk since you were old enough to utter words. Although most toddlers can distinguish between a circle, a square, and a triangle, I doubt many of them could tell you why that knowledge is important. And, fortunately for their parents, they tend to ask "why?" about everything except geometry!

Despite years of singing about, pointing to, and coloring geometric shapes, you've only seen the tip of the iceberg. As you'll soon learn, there's so much more to geometry than identifying shapes. There's even more to it than calculating lengths, areas, and volumes. I don't have to be psychic to know that geometry will be in your future. Even if you never take another math class, concepts from geometry are used in art, chemistry, medicine, physics, astronomy, automobile design, road construction, real estate transactions, and much, much more! If you reach the end of your life and you haven't made use of geometry at least once ... well, then you will have led a sheltered life. But in order to lead a sheltered life, you would need a shelter. And the only way to have a shelter is to use geometry. Gotcha! Let me elaborate on some of the traditional ideas that will be emphasized in this book.

How Does Geometry Measure Up?

Let's face it. Size matters. Chances are that from the day you were born, your height and weight were recorded. Your size was monitored to ensure your healthy development.

You might have also measured the distance you've traveled, the size of your neighbor's television set, and most things that you are insecure about. Here are a few examples of some occasions when you might want to measure:

♦ When people travel from here to there, one of the most frequently asked questions is "How far is it?" Although geometry can help answer that question, it cannot answer the question "Are we there yet?"

♦ A display of diving skills is best post-poned until after the depth of the water is measured, unless you want to risk a massive headache.

Eureka! _____

When you talk about the size of an object, you need to be specific about what you mean. You could be discussing length, area, volume, or some other property of the object.

♦ Property owners usually want to know the size of their land. Then they can determine how long it will take to mow their yard, and whether or not it's worth paying for lawn service!

You also have to come to terms with what people mean by "size." They could be talking about length, area, volume, or a host of other things. But one thing is certain: in our everyday lives there are lots of opportunities to measure.

Getting Into Shape

You are probably familiar with the basic shapes. Heck, a three-year-old is probably familiar with triangles, rectangles, squares, and circles, but that's just the beginning. In addition to the more simple shapes, there are parallelograms, rhombuses, kites, parallelepipeds, spheres, cylinders, hexagons, tetrahedrons, and some that we haven't learned to name yet! Geometry involves the study of all of these shapes. In this book, you will discover the rules used to name these shapes, and you'll also learn how to construct a few of them.

Tangent Line
Shapes have been important to mathematicians for several thousand years. There is even evidence in their epitaphs. For example, Carl Gauss was a famous mathematician in the nineteenth century. He was the first to discover the construction of a heptadeco-gon (a regular 17-sided polygon). There is a rumor that he asked to have one carved on his tombstone.
Archimedes, a famous Greek mathematician, is known for crying out "Eureka!" It is said that he has a sphere inscribed in a cylinder on his tombstone.

Let's Be Reasonable: Learning How to Write Proofs

Geometry might be the first subject in your mathematical history where you will be expected to defend your work. Most math classes tend to be algebraic in nature: The teacher typically gives you problems in the form of a bunch of numbers and symbols, and a set of instructions that involve finding "the answer." You can work the problems in your head, on your fingers, or in the margin of the paper. You do not have to explain yourself, or show your work, much less defend your answers (unless you are trying to get partial credit). But in geometry, the rules are different. You are given the answers (in the form of a theorem) and asked to show how to get that answer. It's a bit of a cheat: You know what the outcome needs to be, and your job is to show how to get there. But you have to defend your every move. That defense of your work is a *proof*: a sequence of statements that starts with a given set of premises and leads to a valid conclusion, with each step supported by a valid reason.

Oh yeah, before I forget. You will not be writing your proofs in a diary or journal. They are not personal or confidential. You should be proud of your proofs. Just as I want everyone to read this book, you will want everyone to read your proofs. And when the casual reader examines your proofs, be prepared for them to ask you "why" at every step. Think of the embarrassment that would result if you couldn't answer them. By the way, I might as well get this out in the open right now. "It's obvious," is not a valid reason. Neither is "Because I said so." That might work for parents, but it doesn't carry much weight here. Sorry, but that goes for you, too, Mom and Dad. I don't play favorites.

> **Solid Facts**
>
> A **proof** in geometry is a sequence of statements, each supported by a reason, that starts with a given set of premises and leads to a valid conclusion.

Who's Idea Was It?

Most of the stories told about the early Greeks focus on either the politicians or the gods. The Greek mathematicians, while leading equally exciting lives, seem to have been omitted from the most famous legendary tales. There are no complete texts of Greek mathematics dating earlier than around 300 B.C.E. In fact, it took hundreds of years before the stories about early Greek mathematicians were written down … just enough time to embellish the facts! But that's how great legends are created.

The Tale of Thales

Thales is credited with beginning the Greek mathematical revolution. He was one of the first mathematicians to support his discoveries with logical reasoning. He developed some particularly useful problem-solving techniques using geometry. One particularly clever idea enabled him to measure the distance from the shore to a ship at sea without swimming to the ship with a tape measure. He is also credited with discovering theorems about isosceles triangles. There are many stories told about him, most of which have been stretched by imagination and time. Rumor has it that his contemporaries held the opinion that he wasted his time on idle pursuits, like the study of geometry! Thales supposedly grew tired of this reputation and decided to take action. One year he noticed signs indicating a bountiful olive crop, and quietly bought most of the local oil presses. When the large crop was harvested, he was the oil press go-to guy, and he made a small fortune. He successfully demonstrated that his pursuits weren't always idle, and that it was possible for a mathematician to exploit a situation and get rich. Most mathematicians today are *not* rich. This is because mathematicians prefer to engage in idle pursuits, like the study of geometry, rather than make their fortunes day trading!

And the Winner Is ... Euclid

The name most often associated with ancient Greek geometry is Euclid. He is generally thought to be the "father" of geometry. This isn't because he is thought to have "sired" geometry (that credit is given to Thales), but because of his strong influence in shaping geometry. He nurtured it as it grew into the fine subject it has become. Euclid helped geometry develop by having the discipline to write down the details. He was the first one willing to admit his assumptions and define his terms, and he proved everything he could along the way (which really was a lot).

His text, the *Elements* of Euclid, consisted of thirteen books, and has appeared in more editions than any work other than the Bible. His book was probably one of those "classics" that everyone had on their shelves, but few bothered to read. There is a reason for this: The original text was not terribly exciting. There were no clever remarks, stories, examples, or calculations. There were, however, plenty of postulates, definitions, theorems, and proofs. You might have heard these terms before; mathematicians tend to throw them around like a pillow at a slumber party. Inspired by Euclid, I'd like to take a minute to formally introduce you to some terms.

A *definition* of a term is a description that captures the essential qualities of the object being described. That seems easy enough, but there is a catch: Everyone has to agree on the "essential qualities." For example, I might think that an essential quality of a dog is that it likes to catch a Frisbee, but a biologist might focus more on the arrangement of its teeth and the shape of its tail. Because a biologist is more of an expert in these matters, I am willing to accept his or her definition of a dog. But I won't let that definition dictate what kind of dog I decide to adopt. My dog must like to catch a Frisbee! Unfortunately, not all words can be defined. For example, let's think about the essential qualities of a point. On a recent survey, the most frequently listed essential quality of a point is "Ummmm ..." That's the point: Nobody's come up with one yet, not even Euclid.

Now that I've shown you some of the difficulties in defining terms, I can move on to constructing belief systems. You might *think* it is easy to state what you believe, but trust me on this: Philosophers and religious leaders have struggled with constructing belief systems for thousands of years. Beliefs are tricky: The more you explore them, the fewer you have. Eventually you reach a point where you end up shouting something like "That's just the way it is! Don't ask me why!" Fortunately, you won't have to build the belief system for geometry, because mathematicians have spent over 2,000 years developing it for you. Mathematicians refer to beliefs as postulates. A *postulate* is a statement that is assumed to be self-evident, and will not (or cannot) be proven.

Armed with undefined terms, defined terms, and postulates, you will prove statements that are called *theorems*. Then you will combine these theorems with undefined terms, defined terms, and postulates, and prove more theorems. And so it goes.

Over time, more and more people became curious about Euclid's volume of work and started reading it more carefully. As more people read it, it became more controversial. It might be a stretch to imagine a controversial math book, but just work with me here. One of the biggest issues that mathematicians have debated over the centuries is Euclid's parallel postulate (also known as Euclid's 5th postulate). As a result of this debate, we now have several geometries: Euclidean, spherical (or elliptic), and hyperbolic. These different geometries were developed in the early eighteenth century by Saccheri, and the early nineteenth century by Gauss, Lobachevsky, Bolyai, and Riemann. Not surprisingly, these results have only recently become appreciated. As a species, we are often slow to embrace new ideas, especially when they are embedded in controversy or mathematics.

> **Solid Facts**
>
> A **definition** is a description that captures the essential qualities of the object you are trying to describe.
>
> A **postulate** is a basic statement that is assumed without proof to be self-evident.
>
> A **theorem** is a statement that can be proven using definitions, postulates, and other theorems.

> **Tangent Line**
>
> Algebraic facts that are accepted as true are usually referred to as axioms, and mathematicians reserve the term postulate for geometric facts that are accepted as true.

Tangent Line

Girolamo Saccheri tried to prove Euclid's 5th postulate by assuming it was false and looking for a contradiction. Despite his "failure," his work eventually led to the development of hyperbolic geometry. Unfortunately, he was such a firm believer in Euclid's 5th postulate that he doubted his results and tried to weasel out of them. If he had published his results without the weaseling, perhaps hyperbolic geometry would now be known as Saccherian geometry.

In the late nineteenth century, a mathematical revolution began. This was a nonviolent revolution, and did not make the national news. During this revolution, rigorous mathematical systems were in vogue. Mathematicians began to develop these systems left and right. This was a drastic change: Before that time mathematicians were not very clear about "simple" things, such as how to define a real number. But this revolution changed all of that. Basic terms were clearly defined, and all assumptions were spelled out as postulates, argued about, and voted on. These discussions about definitions and postulates breathed new life into the mathematical community, and it was an exciting time to be a mathematician. Because mathematicians were new at constructing these rigorous systems, they turned to Euclid's work for guidance. But, much to their dismay, even Euclid's writings fell short of the "new standards." Mathematicians began to complain about Euclid's proofs, especially about some assumptions used in his proofs that were not explicitly mentioned in his list of postulates. Then David Hilbert came along to save the day (or the subject).

David Hilbert: A Goliath Mathematician

David Hilbert has been described as one of the last many-sided mathematicians (I don't mean that he was a polygon, I mean that he was a versatile mathematician). During this mathematical revolution, he took it upon himself to nail down the definitions and postulates of geometry once and for all. He started with three undefined terms: a point, a straight line, and a plane. He developed the relationships between these terms using postulates that were clear and precise. Hilbert's goal was to remove any geometric "intuition" from his results: Every assumption was listed as a postulate. He did not include any statement as an "obvious" fact. The mathematical community was happy, and the foundation of geometry was firm and sound.

Tangent Line

A brief history of geometry: Thales started it, Euclid wrote it down as best he could, and Hilbert polished it and set it on a sturdy foundation.

Can I Really Learn This?

Now that you have an idea of where geometry has been and where we will be going with it, I'm sure you are eager to embark on your journey into the inner and outer limits of geometry. The road to geometry is paved with more than good intentions. It's paved with visualization, investigation, and discovery. Just keep in mind that as you accompany me down the road, you will be propelled by reason.

As confusing as it all might seem right now, geometry is a fairly straightforward subject. I will clearly spell out the rules involved in formulating a valid argument along the way. I want to warn you, though, that writing proofs is not a spectator sport: You'll need to put the "me" in geo**me**try! Although it is important that you are able to follow along and understand the proofs I write, it is also important that you take your turn writing them (that's the "try" in geome**try**). You should feel free to use my proofs as a model, and the solutions in the book as instant feedback. At first, it might be hard to know where or how to start, and it might take some practice learning how much detail to include in a proof. But as you work through the ideas, writing proofs should become second nature.

From my experience as a geometry teacher, I believe that there are five basic steps to being successful in learning geometry:

◆ **Understand the definitions.** English isn't the only subject where you have to learn definitions. To make your life (or at least learning geometry) easier, I will present all important vocabulary terms in simple English, and I promise there will be no spelling tests. The goal here is that you understand what the terms mean and how to make use of them in formulating your arguments. This might involve some memorization. Fortunately, the definitions will make sense, making any necessary memorization easier. And if you need it, the glossary at the end of the book will refresh your memory.

◆ **Read the postulates carefully.** The postulates are your belief system. Take them on faith. They should be reasonable, and might be even intuitively obvious to the most casual passerby. I will translate each of the postulates into plain English and make any underlying implications crystal clear.

◆ **Develop a mathematical instinct.** Some people believe that with geometry, you either get it or you don't. But that's not one of Euclid's postulates, so you don't have to incorporate that idea into your belief system. What you will discover down this road is that most theorems are very cooperative. They want to be proven. They will even tell you how they should be proven. Rather than turn a deaf ear, I will help you learn how to tune in to their subtle whisperings and understand their hidden language. You will develop this skill with practice. And because it's more fun to play than to sit on the bench, there will be plenty of opportunities for you to get into the game!

♦ **Understand the theorems I ask you to prove.** You are building a house of cards, with the postulates and definitions forming the foundation, and theorems reaching up to the stars. You will use established theorems to help prove other theorems. By understanding each step along the way, your house will stand through any hurricane that comes along.

♦ **A picture is worth a thousand words.** In preparing to prove a theorem, it is usually a good idea to have a picture illustrating what the theorem is trying to establish. It translates the abstract ideas of the theorem into concrete terms that can be visualized and discussed. Plus, it gives you an excuse to doodle while reading this book! So be prepared to have pencil and paper handy.

The Least You Need to Know

♦ Geometry involves the study of the properties, measurement, and relationships of points, lines, angles, surfaces, and solids.

♦ Geometry has numerous applications to real life and everyday situations.

♦ Geometry has developed over the past 2,500 years. Euclid was the first to write down the details, and Hilbert was the last.

♦ With time and practice, anyone can be a successful geometry student.

Chapter 2

Let's Do Algebra

In This Chapter

◆ What it means to be equal

◆ How to change equations and maintain balance

◆ What it means to be unequal

◆ How to deal with your problems

There's no need to check the title of this book; it really is a book about geometry. I'm sure you are eager to start right in on learning geometry, especially after that first inspirational chapter. If you were hoping to start a fresh adventure with geometry and leave your "algebraic" baggage behind, then the title of this chapter might cause some initial discomfort. I hope you will find comfort in the fact that, although you will not be able to leave your entire algebraic suitcase behind, you can sort through it and only bring the essential items as you embark on our geometrical travels. By the way, you were wise to select me as your guide. As a seasoned tour guide, I have learned to appreciate the number one rule for traveling: Pack less and take more money.

You probably didn't realize it, but while you were on your algebraic journey you picked up many souvenirs. Maybe you couldn't resist a few linear equations, a miniature quadratic formula, or some tips on factoring. The majority of your souvenirs are priceless. One of the nice things about these souvenirs is that you don't have to declare them while going through customs, as there are no export or import fees. And though they are valuable, you don't have to lock them up to keep them safe; no one can ever steal them from you. Your souvenirs are also durable: they won't break from usage. In fact, the more you use them, the stronger they become. They will certainly last your lifetime.

You might also have some unresolved issues with algebra. That's nothing to be embarrassed about. Most of us have issues of some sort. What you do about them is your choice. Some philosophies encourage you to ignore them, and hope they go away. A more constructive option is to work through your issues. The first step to solving your algebraic problems is to admit that you have them. For those of you who have trouble with this first step, I'll do my part by giving you some of my problems.

Long Lost Relations

As you encounter new things, you often wonder how they relate to things you already know. At a family reunion, you might meet people for the first time and wonder "How are we related?" As you encounter mathematical objects, you also wonder how they relate to things that you already know.

With people, you understand what it means to be related: They share a common ancestor. In mathematics, a *relation* connects two elements of an associated collection of objects. Okay, this might sound a bit confusing. By an "associated collection of objects" I mean things that are to be compared. Keep in mind that the objects in a collection must be similar enough to each other to make some comparison. For example, I could wonder how my mother's cousin's daughter and I are related, but I won't explore how my dog and my goldfish are related.

In this chapter, the "associated collection of objects" will be numbers. In later chapters, the associated collection of objects will be angles, triangles, line segments, or some other geometric figures. But they will never be dogs or fish! In mathematics, the elements in an associated collection of objects can be related in a variety of ways. For example, segments can be related because they are congruent, or they can be related because they are parallel. It all depends on what you want to focus on.

Solid Facts

A **relation** connects two elements of an associated collection of objects.

Properties of Equality

The first relation that you will spend some time with is equality. When two things are equal, you can think of them as being interchangeable. In the cooking world, 12 of something is equal to a dozen. You could go to the store and ask for either 12 eggs or a dozen eggs, and expect to get the same thing. A dozen and 12 are not necessarily interchangeable at a bakery, however. Some bakers are generous and throw in an extra donut when you ask for a dozen. I have tried to talk with my baker about equality, a dozen, and the number 12, but he still throws in the extra donut. Maybe it's his way of showing his appreciation for the thorough lesson in mathematics.

In the mathematical world, when two things are equal, they are always interchangeable, with no exceptions. This is referred to as the *substitution property of equality*. Mathematicians have a neat symbol for equality: =. For example, if you're in the middle of a problem and I tell you that x = 3, then everywhere you see an x you can replace it with the number 3. Also, any time that you see the number 3 you can replace it with the symbol x. You are allowed to freely interchange x and 3 throughout the whole problem. Keep in mind that when you move on to a different problem, all values get reset, and you have to rediscover relationships.

Solid Facts

The **substitution property of equality**: if a = b, then a replaces b in any equation.

Example 1: Given that x = 4 + a and a = 7, find x.

Solution: I will use the substitution property of equality. Because a = 7, everywhere I see an a I can replace it with 7:

x = 4 + a

x = 4 + 7

x = 11

Solving these problems really isn't that difficult. Equality lets you take complicated equations and make them simple. Mathematicians use the word "simplify" to describe the process of turning complicated equations into simple equations.

Now that you have your first relation under your belt, you are ready to explore three special properties that relations might or might not have.

Reflexive, Symmetric, and Transitive Properties

A relation has a *reflexive* property if every element in a collection is related to itself. Suppose the collection of objects consists of people, and two people are related if they share a common ancestor. Is every person related to himself? Of course! Suppose that the collection of associated objects contains numbers, and that these numbers are related by equality. In other words, two numbers are related to each other if they are equal to each other. For example, the number 1 and the number $\frac{2}{2}$ would be related to each other, because they are equal to each other. Does equality have a reflexive property? In other words, is a number equal to itself? Yes! What else would it be equal to? You can write this as $x = x$.

A relation is *symmetric* if, whenever x is related to y, y is related to x. Again, look at the collection of people. If I am related to you, does that mean that you are related to me? Yes! Does equality have a symmetric property? In other words, if the number 1 equals the number $\frac{2}{2}$, does the number $\frac{2}{2}$ equal the number 1? Yes! For equality, we can write this as: If x and y are any two numbers and $x = y$, then $y = x$.

The third property that some relations have is called the *transitive* property. It goes something like this: If x is related to y and y is related to z, then x is related to z. Although the analogy with people can shed some light on this property, it can only take you so far. Genealogy can be fairly complicated: If you look too deeply, you're bound to run into problems, primarily because of marriage, divorce, and re-marriage. Because numbers never divorce (they are constant to the end), let's just explore the transitive property of equality with numbers. If the number 1 equals the number $\frac{2}{2}$, and the number $\frac{2}{2}$ equals the number $\frac{3}{3}$, then the number 1 equals the number $\frac{3}{3}$. In general, if x, y, and z are any numbers such that $x = y$ and $y = z$, then $x = z$.

Solid Facts

Reflexive property: x is related to x.

Symmetric property: if x is related to y, then y is related to x.

Transitive property: if x is related to y and y is related to z, then x is related to z.

Eureka!

Equality is a relation that has three important properties: reflexive, symmetric, and transitive.

The reflexive, symmetric, and transitive properties are pretty important properties. Not only do they help you solve equations, but they will help you justify some of the steps in your geometric proofs. Any relation that has these three properties is called an *equivalence relation*. The only way that a relation earns the status of equivalence relation is for it to have all three properties. It must have all three. Sometimes a relation has two out of the three properties, and a sad tale to tell about why it's missing the third property. In this case, you can sympathize with the relation, but that's about it.

There are no exceptions to rules in mathematics, even for hardship cases. No three properties, no equivalence relation. Equality is your first example of an equivalence relation: as has already been observed, it possesses the three required properties. As you explore other relations throughout this book, you will discover which belong to this elite class and which do not.

Solid Facts

An **equivalence relation** is a relation that has reflexive, symmetric, and transitive properties.

Algebraic Properties of Equality

You can think of equality as harmony and balance: Whatever is to the left of equality is balanced by whatever is to the right. Although mathematicians typically don't mind change, when it occurs they usually like to maintain their balance during the process. There are times when you have some extra stuff you don't need, and you want to give it to some equation. In order to maintain balance, you have to give the same amount of stuff to each side of the equality sign. If you've ever dealt with children, you already

know this principle. Suppose there are two children in balance with some cake, and you have some ice cream to give to them. If you give a scoop of ice cream to one child, and you want to maintain that balance, you'll need to give an identical scoop to the other child. Of course, if you want to experiment, try giving one scoop to just one child. I guarantee you'll only do this once in your life! This notion of maintaining balance when giving stuff away is referred to as the *addition property of equality*: if x = y, then x + z = y + z. Notice that I started with x and y in balance, and then I threw z into the mix. I was able to maintain balance by adding z to both sides. If you add z to just one side, be prepared to hear some screaming!

Solid Facts

The **addition property of equality**: if $x = y$, then $x + z = y + z$.

The **subtraction property of equality**: if $x = y$, then $x - z = y - z$.

The **multiplication property of equality**: if $x = y$, then $x \times z = y \times z$.

The **division property of equality**: if $x = y$ and $x \neq 0$, then $\dfrac{x}{z} = \dfrac{y}{z}$.

Although it might be better to give than to receive, sometimes you have to take things away. Again, if you start out in balance, and then take things away, you have to be fair and take things away from both sides of the equality. This is known as the *subtraction property of equality*: if x = y, then x − z = y − z. Notice that I started with x and y in balance, and then I threw z into the mix. I was able to maintain balance by subtracting z from both sides.

There is more to life than just adding and subtracting, though. Sometimes you multiply. If you're brave, you even divide. Whatever you do, keep in mind the golden rule of mathematics: "Do unto one side as you do to the other." In the egalitarian world of mathematics, if you do one thing to one side, you are obligated to do the same thing to the other side. You can probably predict what the *multiplication property of equality* says: if x = y, then x × z = y × z. Last, but certainly not least, is the *division property of equality*:

If x = y, then $\dfrac{x}{z} = \dfrac{y}{z}$.

Tangled Knot

In order to maintain equality while rearranging an equation, keep in mind that whatever you do to one side of an equation, you must do to the other side.

The only requirement here is that z cannot be zero.

You can solve lots of algebraic equations with these new tools. Remember to play fair, and follow the golden rule. Use only one property at a time, but don't be surprised if you use more than one property to solve the problem completely.

Example 2: Solve for x: 2x + 3 = 11.

Solution: You want to get x all by itself, so get rid of the stuff on the left that doesn't involve x. Right away, I see the number 3 hanging around. It doesn't involve x, so it needs to go! I will use the subtraction property of addition and subtract 3 from both sides and then simplify:

2x + 3 − 3 = 11 − 3

2x + (3 − 3) = (11 − 3)

2x + 0 = 8

2x = 8

That's better. Now I have only one term on the left that involves x, but there's a 2 in front of it. The 2 shouldn't be there, so use the division property of equality and divide both sides of the equation by 2:

$\dfrac{2x}{2} = \dfrac{8}{2}$

x = 4

Whenever you have to solve for x, the first thing you need to do is collect all of the terms that involve x and bring them to one side of the equation using either the addition or subtraction properties of equality. Take all of the terms that don't involve x to the other side of the equation, again using the addition or subtraction properties of equality. Combine the terms on each side, then use either the multiplication or division properties of equality to finish up the problem. Let me do another example before I move on to the next property.

Example 3: Solve for x: 3x – 9 = x + 1.

Solution: Things are getting a little tricky, but it's nothing you can't handle. Notice that x appears on both sides of the equality. Gather all of the x's together using only the addition or subtraction properties of equality. I'll use the subtraction property of equality, and subtract x from both sides:

3x – 9 – x = x + 1 – x

2x – 9 = 1

Next, move all the terms that don't involve x over to the other side of the equality. Use the addition property of equality and add 9 to both sides:

2x – 9 + 9 = 1 + 9

2x = 10

There's just one thing left to do: use the division property of equality to get rid of the 2. Divide both sides of the equality by 2, and simplify:

$$\frac{2x}{2} = \frac{10}{2}$$

x = 5

If you approach these problems the same way each time, you'll turn your problems into solutions in no time!

The Square Root Property of Equality

The square root of p, symbolized by $\sqrt{p}$, represents the number that, when multiplied by itself, equals p. Notice that $\sqrt{p} \times \sqrt{p} = p$. Another way to write this is $\left(\sqrt{p}\right)^2 = p$. You might recall the square root property of equality from algebra. When it is discussed in that setting, the square root of a number is not unique. There are actually two numbers that, when multiplied by itself, equal p: $\sqrt{p}$ and $-\sqrt{p}$ (unless p = 0, of course). In geometry, you will be using the square root property of equality as a tool to solve for lengths of line segments or measures of angles. These lengths and measures can only be positive numbers. So for our purposes, the square root property will give a unique positive answer. Just remember to include both solutions on any future algebraic journey.

The Square Root Property of Equality: Let x represent the length of a line segment, and let p represent a positive number. If $x^2 = p$, then $x = \sqrt{p}$.

Properties of Inequality

I'd like to take this time to formally introduce you to another relation: inequality. You have met inequality before, under its other name "greater than." It also has another side to it: "less than." But once you understand "greater than," "less than" will just involve tilting our heads. "Greater than" has the symbol > associated with it. If x is greater than y, write x > y (and say in your mind x is greater than y). "Less than" has the symbol < associated with it. If you'll tilt your head (or just turn the book upside down), you'll see how the two are related. They're just the opposite of each other. So, the opposite of everything I say about > holds true for < . Just don't read this part of the book on "opposite day," or we'll all be confused!

You might already have a feel for the ideals that > embodies: bigger, more, and so on. However, just to be sure that we are on the same page, I'll explore the ideas in more mathematical terms. If x > y, then x has more than y. There is an imbalance between x and y. If there wasn't an imbalance, then x and y would be balanced, or equal. By the very term "inequality," you sense that they are not equal. So what would you need to do in order to bring balance to the world of x and y? You could either give more to x or more to y. Well, x already belongs to the "have" club; y belongs to the "have-nots." If you want to restore balance, or right the wrong, then it makes sense to give to the "have-nots," which in this case would be y. That's how mathematicians see it, at least.

Now you are in great shape to understand the official definition of > : If x and y are real numbers, x > y if there exists a positive number z for which x = y + z. In other words, you must give something positive to y in order to restore balance.

> **Solid Facts**
>
> If x and y are real numbers, x is **greater than** y (x > y) if there exists a positive number z for which x = y + z.
>
> If x and y are real numbers, x is **less than** y (x < y) if there exists a positive number z for which x + z + y.

Notice that in your definition of > , you don't have to actually know the value of this positive number z, it's enough to know that it exists. Sometimes it is easier to say that something exists than to actually *find* it. This is one of the few places in mathematics where it is okay to take the easy way out.

When you have an official definition, it is good practice to try to use it. A concrete example can usually solidify your understanding of an abstract idea. So I will take a minute and discuss a concrete example that satisfies the definition of >.

It might be common knowledge that 5 > 2. Can you make sense of it using the official definition? The definition of > requires that something positive be added to 2 to restore the balance. What positive number can you add to 2 in order to equal 5? If you add 3 to 2, the balance is restored. In other words, because I can find a number (in this case, 3) to add to 2 (the smaller number) so that the sum balances or equals 5, I know that 5 > 2.

As another example, let's see if 7 > 9. If it were, then I would be able to find a positive number that I can add to 9 in order to get 7. But the only number that I can add to 9 in order to restore the balance is –2, and –2 is not a positive number. So 7 is not greater than 9.

Is It an Equivalence Relation?

Remember that an equivalence relation requires three properties: reflexive, symmetric, and transitive. Let's see if the relation > has what it takes.

First, there is the reflexive property. An element must be related to itself. Well, right away you run into problems. Is a number greater than itself? No. I convinced you just a few pages ago that a number is equal to itself. So > does not possess the reflexive property. It didn't take long to establish that > is not an equivalence relation.

Even though you know that > is not an equivalence relation, it is still worthwhile to determine if it has any of the other desirable properties, like the symmetric or transitive properties. Next on the list is the symmetric property: if x > y, is y > x? Certainly not. So far, this relation is not doing so well. It doesn't have the reflexive or the symmetric properties.

There's one last property to explore: the transitive property. This is the last important property that > could have. The transitive property applied to > reads: if x > y and y > z, then x > z. That one seems to work. If x has more than y which in turn has more than z, then it seems reasonable to conclude that x has more than z.

Eureka!

The relation > is not an equivalence relation. It does not have a reflexive or symmetric property, but it does have a transitive property.

Algebraic Properties of Inequality

You can think of > as imbalance: If something is greater than something else, you need to add to that something else in order to restore balance. You can examine what happens to this imbalance when you have extra stuff to give away. Again, be equitable with what you are giving to each side.

Suppose you start with an imbalance: $x > y$. If you have an extra amount, say z, to donate to both sides, the imbalance will still remain. As long as I'm fair to both sides, and give each side the same amount, it won't matter how generous I am. The gap between the two sides will remain. And that makes sense. If I have \$10 in my pocket and you have \$5 in your pocket, and some kind stranger gives us each \$20, I'll still have more money than you. In symbolic form, you could write: if $x > y$, then $x + z > y + z$. The only way to close the gap is to give to the "have-nots," as I discussed earlier, but mathematicians tend to give equally to both sides. This never-changing gap phenomenon is called the *addition property of inequality*: if $x > y$, then $x + z > y + z$.

> **Solid Facts**
>
> The **addition property of inequality**: if $x > y$, then $x + z > y + z$.
>
> The **subtraction property of inequality**: if $x > y$, then $x - z > y - z$.
>
> The **multiplication property of inequality**: if $x > y$ and $z > 0$, then $x \times z > y \times z$.
>
> The **division property of inequality**: if $x > y$ and $z > 0$, then $\frac{x}{z} > \frac{y}{z}$.

It is also true that mathematicians take things away in a fair manner. If I start with the imbalance $x > y$ and I take an equal amount away from both x and y, the gap between x and y will remain unchanged. Again, the key is that I am taking away the same amount from each side of the inequality. This is known as : if $x > y$, then $x - z > y - z$.

Multiplication and division are a little more tricky. If I multiply or divide by a positive number, then what I expect to happen does happen. The multiplication property of inequality says that if $x > y$ and z is a positive number, then $x \times z > y \times z$. Finally, the division property of inequality says that

if $x > y$ and z is a positive number, then $\frac{x}{z} > \frac{y}{z}$.

> **Tangled Knot**
>
> When using the multiplication or division property of inequality, be sure that z is positive.

I could go into more detail about what happens if z is negative, but on this leg of the journey you need to carry less and bring more money. In any situation encountered in this book, z will always be positive, so there's no need to carry that extra baggage because you won't need it.

An Additional Additive Property

Although I like to think that we mathematicians do unto one side as we do unto another, there are times when we start out with an imbalance, and we add insult to injury by playing favorites. When that happens, as long as we are consistent with our favorites (you must help the winning side more than you help the losing side—I know it's not right, but you can't ruffle the winning side's feathers) we won't run into trouble. In other words, if you have an imbalance, you are only allowed to make the imbalance worse, not better. Let's put things in mathematical terms.

Additive property of inequality: If $a > b$ and $c > d$, then $a + c > b + d$.

Put Me in, Coach!

Here's your chance to shine. Remember that I am with you in spirit and have provided the answers to these questions in Appendix A.

1. Given that $x = a - 7$ and $a = 10$, find x.
2. Solve for x: $4x - 12 = x + 6$.

The Least You Need to Know

 ◆ The three important properties of a relation are the reflexive, symmetric, and transitive properties.

 ◆ An equivalence relation is a relation that has all three of the above properties. Equality is an example of an equivalence relation.

 ◆ The relation inequality is not an equivalence relation: It only has the transitive property.

Chapter 3

Building Blocks

In This Chapter

♦ Learn to speak "geometry"

♦ Models for a point, line, and plane

♦ Meet another relation

♦ Adopt new symbols

Now that you've survived your encounter with algebra, you're ready to travel to the world of geometry. When you travel to distant lands, you should never assume that everyone speaks your language. There are important words and phrases that you should learn in the language of the land in which you travel. Words like "please" and "thank you" will earn you many points when talking with the locals and asking for information or help. Phrases like "Where is the bathroom?" and "How much will it cost?" are also helpful if you need some relief or want to buy such items as food or souvenirs.

As I've already mentioned, Euclid's geometry is based on defined terms, undefined terms, and postulates. These are the building blocks that you will use to demonstrate the truth of hundreds of theorems. I realize that I keep talking about these defined terms, undefined terms, and postulates without giving you any specifics. Which terms are defined? How do you define a word? Which terms are undefined? How do you undefine a word? And what the heck does a postulate look like, anyway? Enough talk! Let's get to the lingo.

Coming to Terms

Definitions help us communicate ideas. They are used to create a vocabulary for describing objects, numbers, concepts, and their relationships. We introduce new words when something important needs to be pointed out. Defining terms requires precision, and vague words like *some* and *small* should be avoided. When you create a definition, you need to do two things: Put your object into a class of other well-defined similar items, and state how it differs from other items in that class (or what special properties it has that other objects in the class do not have). For example, if I wanted to define a dog, I could start by identifying it as a mammal. In doing so, I have put my object (a dog) into a class of other well-defined items (mammals). Then I need to discuss how it is different from all of the other mammals that are already defined, like, say, kangaroos. If there's no difference between a dog and a kangaroo, then there would be no need to invent and define the word "dog;" the word "kangaroo" would work just fine.

With the English language, we often get bored with using the same word over and over again. Synonyms are words that mean close to the same thing, like the words *help* and *assist*. English is rich in variety. But in mathematics, variety is not the spice of life, it is the bane of our existence. What's the point of spending days thinking up a word, like "obtuse," and coming up with a good definition for it, if the next week we have to come up with another word that means the same thing? It is difficult enough to come up with one word to convey a particular idea. Mathematicians can use the same words over and over again without any risk of boredom. We prefer to conserve our resources and keep the word definitions to a minimum. That's an advantage to those of you trying to learn these new words. If there are fewer words defined, then there are fewer definitions to learn. That will keep the glossary at the end of the book relatively short, which means less work for me! So we all win!

Points and Space

A *point* is the basic unit of geometry. You can't see it. It has no size. It is infinitely small. There are three important features about a point: location, location, and location! As with most new ideas, you need to find a physical model for a point, something to help you visualize what a point is. People often use physical models to gain insight into an idea, even though the model might be technically incorrect. You can use a dot to represent a point, even though a point is smaller than the smallest dot you can draw. For consistency, label your points with a capital letter. The point shown in Figure 3.1 is called P. If you were to draw points the way they

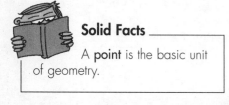

Solid Facts

A **point** is the basic unit of geometry.

really are, your sightseeing would be nonexistent, and your tour would be over before it began. Euclid tried to define a point as "that which has no part." That could just as easily refer to an out-of-work actor as a nonexistent dot on a page.

•
P

Figure 3.1

The model of a point P.

Tangled Knot

You can name your points anything you want. The main reason for labeling points is so that you can refer to them easily. You could call your point **P** or **A** or any other letter of the alphabet. But be warned! After you label a point, you are not allowed to change its name, even if it gets married! So don't give your point a goofy name that it will resent, or that you'll regret later. And don't name them all *George!*

Accept "point" as your first undefined term. You will learn to "undefine" your terms by describing them as best you can, and accept your inability to write down an exact definition. You probably have an intuitive feeling about these undefined terms, not that you need it. You just have to start somewhere, and points are as good of a place to start as anywhere else.

Everything that exists in the land of geometry is made up of points. Because points have a location, but no size, we know that they have to live somewhere. That somewhere is *space*. Some people would define space as "the final frontier." But that definition doesn't exactly help on this particular trek. Space is another one of those terms that is difficult to nail down. Mathematicians define space as the set of all points. Of course, I haven't defined *set*. In fact, set is one of those undefined terms. Think of a set as a kind of bag that holds the things you are concerned with. The reason that a set cannot be defined a set is that we don't know what the bag is made out of. If everything in our world is made up of points, then this bag must be made up of points. But then, if this bag contains all points, our bag must be inside of our bag, which is kind of weird. That's why we can't really talk about the bag. The important thing, though, is that there's nothing beyond space.

Solid Facts

Space is the set of all points.

Lines

A *line* is a straight arrangement of points. "Straight" is another one of those words that can't be defined. Mathematicians have an intuitive idea of what it means to be straight: no curves, wiggles, or bends. I'd like to point out some of the mind-boggling properties of a line. First of all, a line contains infinitely many points. It has infinite length but no thickness. It extends forever in two directions. Lines have infinite length,

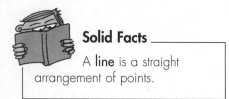

Solid Facts

A **line** is a straight arrangement of points.

yet are made up of points, which have no size. It is bizarre that if you line up *enough* things with no size, you can get something that has infinite size. A physical model for a line is a thin wire. A line, though, is thinner, longer, and straighter than any wire ever made.

If you had to draw a line in its entirety, you would never finish your task. That is because a line extends forever in both directions. Whenever you draw a figure that should represent a line, you can only draw part of it (and you should only draw the part that you want to talk about). Put some arrows on each end to indicate that, although it goes on forever in both directions, you grew tired of its never-ending saga.

You name a line by specifying two points on it. The line in Figure 3.2 is called the line AB. You could also call it the line BA. It doesn't matter which of these names you choose, because a line goes on forever in both directions. I will use the symbol ↔ when I am discussing a line. I will put this symbol on top of the points to indicate that I am talking about the line containing those specific points. The line in Figure 3.2 can be written $\overleftrightarrow{AB}$ or $\overleftrightarrow{BA}$. Because there are infinitely many points on a line, there are infinitely many names for a line. I would run out of letters before I ran out of points on a line. To keep your interest, I will use lots of letters to name points on a line. Sometimes I will call the points A and B, sometimes P and Q, and just to mix things around I'll even use S and T.

Figure 3.2

The line AB (a.k.a. $\overleftrightarrow{AB}$).

Now that you have a description of a line, there are some facts you need to get straight. Because lines are made up of points, it is not surprising that lines and points are closely related. In fact, every line contains at least two distinct points. A line is actually made up of infinitely many points, but because I haven't talked much about infinity, you might be less inclined to believe me about that one. It seems more reasonable to believe that, given any line, you can find two different points that lie on that line. Otherwise, all the points on the line would occupy the same spot on the paper, and we wouldn't have much of a line to speak of. It would certainly have a hard time going off forever in both directions! This is what our first postulate is all about.

Postulate 3.1: Every line contains at least two distinct points.

What you gain from this postulate is that any time you are given a line, you can pluck off a couple of points and label them. (Remember: You are allowed to decide the names of your points, but after you have decided on a label for a particular point, you can't change your mind.) You might wonder how that will help you on your travels. Well, there are lots of times when you will be given lines when what you really need are points. This postulate gives you your points.

Sometimes you might find yourself in the opposite situation: You are given points when what you need is a line. When that happens, you will want to use the following postulate.

Postulate 3.2: Two points are contained in one and only one line.

You can gain some insight into a line by examining Postulate 3.2 a little more closely. If you draw two points and try and connect them using two different lines, one of the lines (probably the second one you try to draw) must have some sort of bend in it (to make it different than the "straight" one that you undoubtedly drew for the first one). This will not do; a line has no "bendy" parts.

Because lines contain infinitely many points, it is not unusual to need to talk about three or more points that lie on the same line. Because it happens so frequently, mathematicians have a special name for it. When three or more points lie on the same line, we call them *collinear points*. On the flip side, points are *noncollinear* if there is no line that contains all of the points.

Solid Facts

Collinear points are points that lie on the same line.

Noncollinear points are points arranged so that there is no line that contains all of the points.

Line Segments

There are times when you are given two points on a line, and you only want to deal with that part of the line between those two points. For example, in Figure 3.3 you might only be concerned with points A and B, and what is in between them. You don't care that the line goes on forever in both directions. You want to chop off its ends and only deal with the middle. You are allowed to do that with immunity. The line won't take it personally or exact revenge on your family. But after you chop off the ends of a line, you no longer have a line. You are left with only part of the line: a line segment. A *line segment* refers to the two chop-off points A and B and all the points on $\overleftrightarrow{AB}$ between them. The points A and B are called the *endpoints* of the line segment. These two endpoints are crucial in specifying the line segment: they let you know where the segment begins and ends.

Figure 3.3

The line segment AB.

A   B

Solid Facts

A **line segment** AB consists of the two points, **A** and **B**, and all the points between them that lie on the line containing **A** and **B**.

The **endpoints** of a line segment are the two points where the line segment begins and ends.

The **length** of a line segment is the distance between its endpoints.

When naming a line segment, the only things you must specify are endpoints. The line segment that starts at A and ends at B is called the *line segment* AB. It could also be referred to as the line segment BA. Because a line segment is different than a line, you need a new symbol to represent a line segment. And because the difference between a line segment and a line is that a line goes forever (which is indicated by the little arrows) and a segment has a definite starting and stopping point, I will write $\overline{AB}$ when I mean the line segment AB.

Segment Length

Lines are infinite, segments are not. So it is meaningful to ask about the length of a line segment. The length, or measure, of a line segment is the distance between its endpoints. If you are measuring a line segment $\overline{AB}$, denote its length by AB (no fancy symbol above the points, just the two endpoints). There are some not-so-surprising facts about distance that I should mention. The distance between a point and itself is 0, and the distance between any two distinct points must be positive. Because the distance between here and there is the same as the distance from there to here, it is not surprising that AB = BA. These statements are the kind of thing that postulates are made of.

Postulate 3.3: For all points A and B, AB = BA.

Postulate 3.4: The Ruler Postulate. The measure of any line segment is a unique positive number.

The instrument used to measure a line segment is a scaled straightedge, which is a fancy name for a ruler. To measure the length of a segment, place the "0-point" of the ruler at one endpoint of the line segment and define the numerical length as the number at the other endpoint. The units for length depend on the units on our ruler. We could be working with units of inches, meters, or miles, just to name a few.

Tangled Knot _____

I am introducing a lot of notation, and it is important to keep it all straight: $\overleftrightarrow{AB}$ refers to the line, $\overline{AB}$ refers to the segment, and AB denotes its length.

A New Relation

Now that you can measure the length of a segment, it is time to explore how segments can be related. You have to be careful when you relate two segments. You can't talk about two segments $\overline{AB}$ and $\overline{CD}$ being equal. The segments $\overline{AB}$ and $\overline{CD}$ have different endpoints (remember: you are not allowed to change the names of the points!), so they can't be the same. But two segments _can_ have the same length. I will define a relation between segments, which I will call _segment congruence_. This new relation will be based on the old, familiar relation equality. Two segments are said to be _congruent_ if they have the same length. This definition can be used both ways: if $\overline{AB}$ and $\overline{CD}$ are congruent, then AB = CD, and if $\overline{AB}$ and $\overline{CD}$ are two segments with AB = CD, then $\overline{AB}$ and $\overline{CD}$ are congruent. A new relation begs for a new symbol. Because congruence is closely related to equality, I will use the symbol ≅ to denote congruence. If $\overline{AB}$ and $\overline{CD}$ are congruent, write $\overline{AB} \cong \overline{CD}$.

With this new relation you can compare segments based on length. This is a good time to review the special properties that relations can have: the reflexive, symmetric, and transitive properties. Let's see which properties ≅ possesses. Because ≅ is closely related to equality, and equality possesses all three of the coveted properties, you might expect that ≅ also possesses all three properties. In this case, your intuition is right on. You will use the properties of equality to establish the reflexive, symmetric, and transitive properties of ≅. That's how you construct things in geometry: You build on already established ideas that you understand.

Solid Facts _____

Congruent segments are two segments that have the same length.

Tangled Knot _____

Be careful when you compare two line segments: Use the symbol ≅ when comparing the segments, and use = when comparing the segment lengths. We will either write $\overline{AB} \cong \overline{CD}$ or AB=CD.

The first property to examine is the reflexive property: Is a segment congruent to itself? Well, a segment will have the same length as itself, so the answer is "Yes!" Next on the list is the symmetric property: if $\overline{AB} \cong \overline{CD}$, is $\overline{CD} \cong \overline{AB}$? Well, if $\overline{AB} \cong \overline{CD}$, then AB = CD, and because equality has the symmetric property, CD = AB, so $\overline{CD} \cong \overline{AB}$. So $\cong$ has the symmetric property. The last property on the list is the transitive property: if $\overline{AB} \cong \overline{CD}$ and $\overline{CD} \cong \overline{EF}$, is $\overline{AB} \cong \overline{EF}$? Again, you need to put things in terms of equality: if $\overline{AB} \cong \overline{CD}$, then AB =CD. If $\overline{CD} \cong \overline{EF}$, then CD = EF. So by the transitive property of equality, AB =EF, which means that $\overline{AB} \cong \overline{EF}$. Because $\cong$ has all three special properties, it is an equivalence relation.

Segment Addition

Now that you know how to measure segments, you can put two of them together and measure the length of the new segment formed. In order to do this, you will rely on another postulate.

> **Postulate 3.5:** The Segment Addition Postulate. If X is a point on $\overline{AB}$, then AX + XB = AB.

Because a picture is worth a thousand words, let's look at Figure 3.4. If you don't believe this postulate, test it with a model. Take a piece of string and measure its length. Then cut the string into two pieces and measure each piece. If you add up the lengths of the two pieces of string, the result should be the length of the original, uncut string.

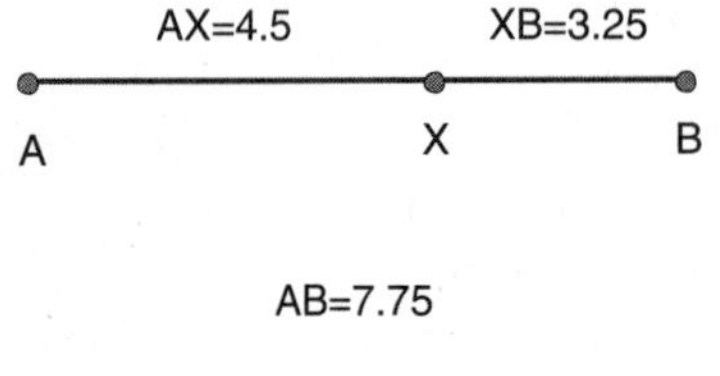

Figure 3.4

AX = 4.5, XB = 3.25, *and* AB = 7.75. *Notice that* 4.5 + 3.25 = 7.75.

AX=4.5   XB=3.25

A   X   B

AB=7.75

Between

"Between" is one of those words that is difficult to define but easy to demonstrate with a picture. Figure 3.5 shows three collinear points, A, B, and C. I could ask most toddlers to identify which point was between the other two, and they would at least be able to point to the letter B. But how do we know when a point is between two other points? Visually, the point that is in between the others has one of the other points to the left and another one to the right. But this is difficult to describe mathematically. Fortunately, you can use the notion of segment length to help you out. Look at the lengths of the different segments involved: AB, BC, and AC. The segment with the biggest length is $\overline{AB}$, and that's the key. Of the three collinear points involved, the one that is between the other two is the one that is not an endpoint of the longest segment. You can define what it means to be between mathematically by using use the Segment Addition Postulate: For any three collinear points A, B and C, B is *between* A and C if AB + BC =AC.

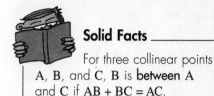

Solid Facts

For three collinear points A, B, and C, B is **between** A and C if AB + BC = AC.

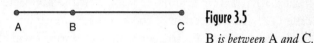

A   B   C

Figure 3.5

B *is between* A *and* C.

There will come a time when you want to divide $\overline{AB}$ into two congruent pieces. In order to do this, find a point M between A and B so that $\overline{AM} \cong \overline{MB}$, or so that AM = MB, as shown in Figure 3.6. The point M that breaks $\overline{AB}$ into two congruent pieces is called the *midpoint* of $\overline{AB}$.

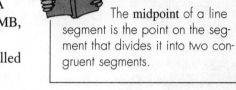

Solid Facts

The midpoint of a line segment is the point on the segment that divides it into two congruent segments.

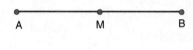

A   M   B

Figure 3.6

M *divides* $\overline{AB}$ *into two congruent segments:* $\overline{AM}$ *and* $\overline{MB}$.

Rays

A ray begins at a point and goes on forever in one direction. You can think of a ray as half of a line; the line has been severed in one direction. A laser beam is a good physical model of a ray. The laser beam has a definite starting point, and has the potential to continue on forever in one direction. The starting point of the ray is called the endpoint of the ray. There are well-established guidelines for naming rays. I have drawn a ray in Figure 3.7. The endpoint of the ray is A, and B is another point on that ray. I will refer to that ray as the ray AB. The order of the points is important. Because I am very concerned with where the line was hacked off, I need to list that point first. Then I need to include the name of another point on the surviving half of the line; in this case it would be B.

Figure 3.7

A ray with endpoint A *passing through points* B *and* Y.

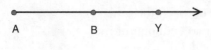

I have already provided you with a symbol for lines (↔) and one for line segments (—). It seems only fitting that to have also a spiffy symbol for a ray. Because a ray is half of a line, I will use half of the symbol for a line: →. I will call the ray shown in Figure 3.7 $\overrightarrow{AB}$. If Y is another point that the ray passes through, the ray can also be called $\overrightarrow{AY}$. The direction of the arrow is important: It indicates which half of the line has been chopped off. To keep things uniform, I will always draw an arrow with the tail to the left and the head to the right. Also, the order in which I identify the points is important. The first point must be the endpoint of the ray. If I called the ray in Figure 3.7 $\overrightarrow{BA}$ I would confuse a lot of people. The abbreviation $\overrightarrow{BA}$ describes a ray with endpoint B that travels off in the direction of A. The ray in Figure 3.7 does nothing of the kind. It is a ray with endpoint A that travels off in the direction of B. You might think that's a minor detail, but think about it in terms of traveling. Starting off at work and traveling past your house as you drive into the sunset is very different than starting off at home and traveling past your work and driving into the sunrise. Unless, of course, you work in your house and you live at your work, in which case you're just spinning your wheels!

Even though I have chopped a line in half to make a ray, a ray is infinite in length. That's because a ray still goes on forever in one direction. For something to have infinite size, all it takes is for it to go on forever in at least one direction.

> **CAUTION**
>
> **Tangled Knot**
>
> With lines and line segments, the order of the points is not important: $\overleftrightarrow{AB}$ and $\overleftrightarrow{BA}$ represent the same line, and $\overline{AB}$ and $\overline{BA}$ represent the same line segment. You must be more careful with rays: $\overrightarrow{AB}$ is not the same as $\overrightarrow{BA}$.

Planes

Airplanes have nothing to do with geometric planes. A plane is another one of those undefined terms. You can describe it, but you can't define it precisely. A *plane* is two-dimensional: It has length and width but no thickness. It is a flat surface that extends forever. A wall, a floor, or a ceiling is a physical model of a plane, but even the wall of the longest building is nothing compared to a plane. A runway at an airport can also serve as a model of a plane, so a plane can land on a model of a plane. You can draw a plane as a four-sided figure, like a piece of paper drawn in perspective. Of course, this represents only part of a plane, as a plane is infinite in length and width.

There is a connection between the "straightness" of a line and the "flatness" of a plane. Lines are as straight as planes are flat. Imagine a plane and two points in that plane. Now think of the line containing those two points. That line must lie completely in that plane. The only way for it to get out of the plane is if it bent up or down. But lines can't bend! So the line is stuck in the plane. This idea is summarized in the following postulate.

> **Postulate 3.6:** If two points are in a plane, then the line containing these points is also in the plane.

Just as two points determine a unique line (Postulate 3.2), three points determine a unique plane. That idea leads to the following two postulates. In fact, the next two postulates relate to planes in the same way that Postulates 3.1 and 3.2 relate to lines.

> **Postulate 3.7:** Every plane contains at least three noncollinear points.

> **Postulate 3.8:** Three noncollinear points are contained in one and only one plane.

Tangent Line

It is common for a chair with four legs to wobble on the floor. This is because it is difficult to get the ends of all four legs to lie in the same plane (the floor). A tripod, on the other hand, is an extremely stable place to rest your camera. A tripod only has three legs, and the ends of the legs are noncollinear. Thus the three legs of a tripod must lie in the same plane (the floor) and there is no wobble. Even if the floor is uneven, the tripod will remain steady.

Notice the similarities between Postulate 3.7 and Postulate 3.1. Both deal with the minimum number of points needed to create a specific geometric figure: two points for a line and three points for a plane. Postulate 3.8 and Postulate 3.2 are also very similar. Both address the fact that the geometric figure determined by the specific number and type of points is unique. Postulate 3.8 does contain a subtle idea, hidden in the word noncollinear. If the word noncollinear had not been included, you might have the situation shown in Figure 3.8. Notice that there are two planes that contain the three points indicated. A great deal of thought went into constructing these postulates, so that they hold up under any amount of scrutiny. As you read these postulates, pay attention to the word choices. After thousands of years and many eyes looking them over, these postulates are in top shape. They are descriptive enough without being overly wordy. All of the words involved have been chosen for a reason. You will gain valuable insight into geometry by thinking about why certain terms were included in a definition or postulate.

Figure 3.8

These two distinct planes contain the same three collinear points.

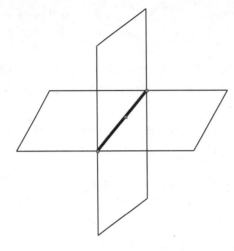

Solid Facts

Coplanar points are points that lie in the same plane.

Noncoplanar points are points arranged so that there is no plane that contains all of the points.

Mathematicians characterize points depending on how they relate to a plane. Points are called *coplanar* if there is a plane containing all of the points, and *noncoplanar* if there is no plane that contains all of the points. Because any three noncollinear points determine a unique plane, you would need at least 4 noncollinear points (in the right arrangement) to form a collection of noncoplanar points. In this book, points shown in figures are assumed to be coplanar unless otherwise stated.

The Least You Need to Know

♦ The three undefined terms are point, line, and plane.

♦ Every line contains at least two distinct points, and given any two points, there is a unique line that passes through them.

♦ Segment congruence is an equivalence relation.

♦ Every plane contains at least three noncollinear points, and given any three noncollinear points, there is a unique plane that contains them.

Chapter 4

There's Always an Angle

In This Chapter

- ◆ Learn the language of angles
- ◆ Add and subtract angles
- ◆ Discover relationships between angles
- ◆ Develop your proof-writing skills

Now that I've introduced you to the basic geometric building blocks, you can use them to build other geometric figures. The first figure you will create is an angle. Angles can be formed in several ways: Two lines, segments, or rays can join together to make angles. You might not notice angles at first, but if you look around, they will start jumping out at you. Look at a door: The top of the door and the side of the door join together to form an angle. As you look around even more, you will probably begin to notice angles everywhere, from the way the walls meet the ceiling to the peak of your roof.

Now that I have you seeing angles everywhere, the time is right to begin our excursion into the world of angles. In this chapter, you will explore these friendly shapes.

What's in a Name?

When two rays (say $\overrightarrow{AB}$ and $\overrightarrow{AC}$) share a common endpoint (in this case, point A), an angle is formed. Actually, two angles are formed, as shown in Figure 4.1, but mathematicians usually only talk about the smaller one. The larger angle is called the *reflex angle*. If they are both the same size, then you can talk about either one.

Figure 4.1

The formation of an angle.

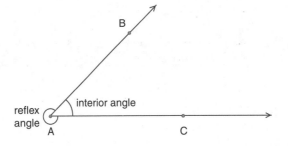

Solid Facts

When two rays share a common endpoint, two angles are formed. The **reflex angle** is the larger of the two angles.

When talking about angles, you should refer to them by name. The name should be unique to the angle in question, so it's not confused with some other angle that's hanging around. You also want this name to be as descriptive as possible. When referring to an angle by name, you want the whole world to know that you are talking about an angle, and not a line segment, a rectangle, or anything else! Use the symbol $\angle$ when you are referring to an angle.

Because an angle is formed by the joining of two rays, it makes sense to give credit to each of the rays for their contribution in forming this angle. Also, it is important to celebrate where the two rays meet (the point A). So, when you name an angle, you will use three points, and angles will have a first, middle, and last "initial." The first initial will denote a point located only on the first ray, $\overrightarrow{AB}$, and not on the other ray, $\overrightarrow{AC}$. It can be any point on $\overrightarrow{AB}$, except for the point A (because point A can also be found on $\overrightarrow{AC}$). Because of the naming convention for rays, you can immediately identify two points that lie on $\overrightarrow{AB}$: points A and B. Use the point B for your angle's first initial. For the second initial, use the common endpoint of the two rays. Because the rays meet at point A, your angle's second initial would be A. The third initial will denote a point that is located only on the other ray, $\overrightarrow{AC}$, and not on the first ray, $\overrightarrow{AB}$. Again, by carefully naming your second ray $\overrightarrow{AC}$, you can immediately identify a point on this ray that is distinct from A: point C. The angle will then be called $\angle BAC$. You might be wondering if it matters which ray to take first. Well, angles are pretty easy-going, and ∠BAC also answers to the name ∠CAB. The only thing that angles care about is that the intersection of the rays is used as the middle initial.

Over time, however, you will probably get tired of writing the angle's full name. So, in addition to an angle's formal name, you can also give it a nickname. Besides the symbol for an angle, think about what other information should be included when referring to an angle. If you ask any happily-together couple what is most important about the formation of their union, they will answer "The place where we met!" Maybe that's what angles think about their formation as well. So your nickname will need only the symbol ∠ and the common point A. Therefore, you will refer to angle BAC as ∠A . Only use nicknames when there's no confusion about which angle you're talking about. If there are a lot of angles in the picture, be sure to use their formal names.

Now that you've named your angle, you can dissect it, and name its parts. The rays $\overrightarrow{AB}$ and $\overrightarrow{AC}$ are known as the *sides* of the angle. The common endpoint of these rays, point A, is called the *vertex* of the angle. Angles have an interior and an exterior, and convention dictates which one goes where. Let's take a look at ∠A in Figure 4.2. Points like D are in the *interior of* ∠A . Points like E are said to be *on* ∠A , and points like F are in the *exterior of* ∠A .

Solid Facts _____

The **sides** of an angle are the two rays that meet to form the angle.

The common endpoint of the two rays is the **vertex** of the angle.

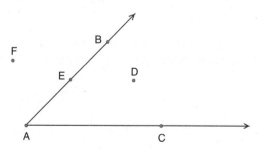

Figure 4.2

Point D is in the interior of ∠A , *Point E is on* ∠A , *and Point F is in the exterior of* ∠A .

Sometimes, you might have a picture that is knee deep in angles. Keeping track of all the angles might become difficult. When this happens, change your notation to "clean up" your drawing and label your angles individually. In this situation, don't bother to label all of the angles, just the ones you plan to talk about.

Are You My Type? The Basic Angle Classifications

Armed with your protractors, you can measure the sizes of your angles. Remember that when two rays meet to form an angle, you only talk about the smaller of the two angles formed. If the two angles are the same size, then you can talk about either one.

As you go about measuring angles, there is one thing you need to take on faith: The measure of an angle is a unique positive number. This statement seems reasonable, and we accept it as a postulate:

> **Postulate 4.1:** The Protractor Postulate. The measure of an angle is a unique positive number.

For the purposes of this chapter, the measures of angles are between 0° and 180°, including 180°. When I discuss trigonometry I will extend the notion of an angle. This will enable you to interpret angles with negative measures or measures greater than 180°, but you will have to wait until the end of Chapter 20 for that. In the meantime, let's review how to measure an angle with your protractor. Place the center mark of the protractor on the vertex of the angle. Rotate the zero-edge of the protractor so that it lines up with one side of the angle. Read the measure of the angle where the other side of the angle crosses the protractor's scale. If the other side of the angle is too short to cross the protractor's scale, extend that side of the angle and try again. Now that the angle has a name and a measurement associated with it, you need an easy way to distinguish between ∠ABC (which is the actual angle) and the measure of ∠ABC, which is a positive number between 0 and 180. Use the notation m∠ABC to represent the *measure of ∠ABC*.

Mathematicians classify angles according to their size. Figure 4.3 shows several angles.

Figure 4.3

Different angles: an acute angle (top left), a right angle (top right), an obtuse angle (bottom left), and a straight angle (bottom right).

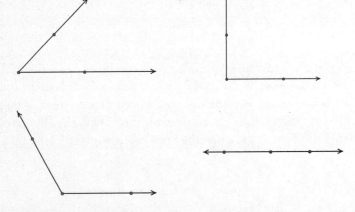

As mentioned in the Protractor Postulate, the measure of an angle must be greater than zero. The largest measure that an angle can have is 180°. When categorizing angles, compare the measure of an angle to 90°, which is half of the maximum measure of an angle (180°). An angle whose measure is less than 90° is called an *acute* angle (not necessarily a cute angle, but it could be). If an angle's measure is exactly 90°, then it is a *right* angle. An angle whose measure is between 90° and 180° is *obtuse*. Finally, if an angle's measure is exactly 180°, it is called a *straight* angle. We can characterize a straight angle as an angle whose sides form opposite rays.

Solid Facts

If an angle's measure is between 0° and 90°, it is an **acute angle**.

If an angle's measure is exactly 90°, it is called a **right angle**.

If an angle's measure is between 90° and 180°, it is an **obtuse angle**.

If an angle's measure is exactly 180°, it is called a **straight angle**.

Angle Addition

Now that you know how to measure angles, it's time to put two of them together and measure the new angle that is formed. In order to do this, you need some guidance. Fortunately, another postulate is here to save the day!

> **Postulate 4.2:** The Angle Addition Postulate. If a point D lies in the interior of $\angle ABC$, then $m\angle ABD + m\angle DBC = m\angle ABC$.

Let's look at Figure 4.4 to see what is going on.

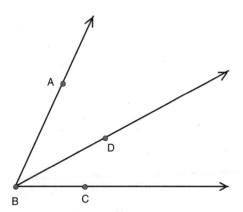

Figure 4.4

To illustrate the angle addition postulate, $m\angle ABD = 34.45°$, $m\angle DBC = 29.30°$, *and* $m\angle ABC = 64.75°$. *Observe that* $34.45° + 29.30° = 64.75°$.

How Do Angles Relate? Classifying Pairs of Angles

At this point, you can measure an angle and classify it as either acute, right, obtuse, or straight. An angle can only fall into one of these categories. It would be a very weird world if an angle were both obtuse and right! You can also arrange angles according to their size. For example, an acute angle always has a smaller measure than an obtuse angle. That's because an angle that measures less than 90° must be smaller than an angle whose measure is larger than 90°. In fact, when it comes to angles, the following pecking order is always observed (starting from the bottom of the pile): acute, right, obtuse, and straight. But there are other important relationships between angles that you will want to explore (and make use of in our proofs down the road).

Adjacent Angles

When I first introduced you to the Angle Addition Postulate, I showed you a picture of three angles (Figure 4.4): $\angle ABD$, $\angle DBC$, and $\angle ABC$. There are some special relationships between these three angles that are worth pointing out. The first relationship that I have already noted was a result of the Angle Addition Postulate: $m\angle ABD + m\angle DBC = m\angle ABC$.

Another relationship that I would like to point out is that all three angles have the same middle initial. They all have point B as a vertex! Whenever that happens, some interesting things can result. Be careful not to jump to any conclusions, though. Although it might seem as if our mathematical world is a pretty exciting place (and that interesting things happen every time you turn the page), there are times when a situation can be a letdown. Let me give you some examples of the exciting, potentially exciting, and disappointing events that might occur.

◆ **Exciting event:** $\angle ABD$ and $\angle DBC$ share a common side and a common vertex. They also have no interior points in common. When this happens, the angles are *adjacent*. That's what happens in Figure 4.4!

Solid Facts

Adjacent angles are two angles that share a common side and a common vertex, but have no interior points in common.

◆ **Potentially exciting event:** $\angle ABD$ and $\angle ABC$ share a common side and a common vertex. They also have some interior points in common. When this happens, we can at least compare $m\angle ABD$ and $m\angle ABC$: $m\angle ABD < m\angle ABC$. This has the potential to be useful and exciting. This is also illustrated in Figure 4.4.

♦ **Disappointing event:** In Figure 4.5, ∠ABC and ∠DBE share a common vertex, but not a common side. Also notice that neither ∠ABE nor ∠DBC are straight. This situation will probably not increase your heart rate. Unless you need some down time, I would recommend moving on to the next exciting situation.

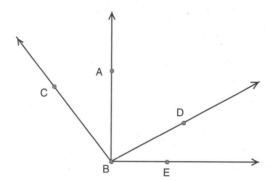

Figure 4.5

A disappointing situation.

Congruent Angles

Sometimes, when you compare two angles, you discover that they have the same measure. This is like finding someone who shares your birthday. An instant bond develops. When this happens in geometry, it is said that the two angles are *congruent*. Congruent angles must coincide when one is placed over the other. Don't worry if the sides of the angles are different lengths. Remember that rays make up the sides of angles, and they are infinite in length. You can always make them bigger or smaller, depending on your needs. If only you could do that with your bank account balances! When you declare two angles to be congruent, the two angles "open up" the same amount.

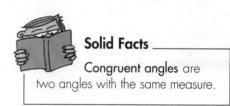

Solid Facts

Congruent angles are two angles with the same measure.

When you have an angle, it's sometimes fun to cut it up into smaller pieces. For example, you can bisect an angle into two smaller, congruent angles. But don't worry, no angles will be hurt in the process. In order to bisect ∠ABC in Figure 4.6, you need to find a point P in the interior of the angle, and construct a ray $\overrightarrow{BP}$ that will divide ∠ABC into two congruent angles. In other words, you need to find a Point P so that ∠ABP ≅ ∠PBC. When that happens, $\overrightarrow{BP}$ *bisects* ∠ABC, or $\overrightarrow{BP}$ is the *bisector* of ∠ABC.

Solid Facts

The **bisector** of an angle is a ray that divides the angle into two congruent angles.

A ray **bisects** an angle if that ray divides the angle into two congruent angles.

Figure 4.6

The ray $\overrightarrow{BP}$ bisects $\angle ABC$ into two smaller, congruent angles: $\angle ABP$ and $\angle PBC$.

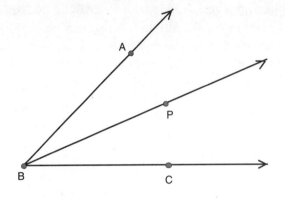

Complementary and Supplementary Angles

You've already observed that angles that measure 90° or 180° are special, and are given their own classifications. Although acute and obtuse angles have measurements that miss these marks, they can join forces with another angle and hit the bull's-eye. When two angles combine so that the sum of their measurements is 90°, the two angles are *complementary*, or each angle is the *complement* of the other. When two angles get together and their angle measurements add up to 180°, the two angles are *supplementary*, or each angle is the *supplement* of the other.

You might think that obtuse angles have issues with self-esteem, because they never get complements, but in reality they are content in the knowledge that somewhere out there is an angle waiting to supplement them. Of course, their options are limited: They can only have acute supplementary angles. Right angles need no complement, and can supplement themselves. Straight angles have no complement, and need no supplement. Acute angles, on the other hand, have more options available. They can find an obtuse angle to supplement them, or another acute angle to complement them. But acute angles don't brag about their many options. You can learn a lot about an angle by knowing its complement or its supplement, as I will demonstrate in the next section.

> **Solid Facts**
>
> **Complementary angles** are two angles whose sum is 90°. Each angle is the **complement** of the other.
>
> **Supplementary angles** are two angles whose sum is 180°. Each angle is the **supplement** of the other.

Algebraic Games We Can Play

You can combine these angle relationships with your freshly toned algebraic skills, and calculate angle measurements.

Example 1: Suppose two angles, $\angle P$ and $\angle Q$ are complementary, and that $m\angle P = \dfrac{x}{2}$ and $m\angle Q = 2x$. Find x, $m\angle P$ and $m\angle Q$.

Solution: Although your tendency might be to just dive right in and start writing equations, take a minute to organize your thoughts. That way, you can practice developing a game plan. At first glance this problem might not seem to warrant all the work involved in creating a game plan, but if you establish good problem-solving skills now, when the problems are small, you will be better equipped to handle the larger problems down the road.

Here's the game plan: Start with each piece of information given. You are told that $\angle P$ and $\angle Q$ are complementary. Right away, the wheels are turning. What do you know about complementary angles? Their angle measurements add up to 90°! Let's draw a picture to visualize things better (see Figure 4.7).

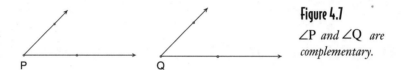

Figure 4.7

$\angle P$ and $\angle Q$ are complementary.

Okay, so you know that

$$m\angle P + m\angle Q = 90°$$

You are given that $m\angle P = \dfrac{x}{2}$ and $m\angle Q = 2x$. Incorporating this information into the above equation yields:

$$\frac{x}{2} + 2x = 90°$$

$$\frac{5x}{2} = 90°$$

$$x = \frac{2}{5}(90°) = 36°$$

So, now you know that $x=36°$. After you solve for x, there's a tendency to move on to the next problem. However, it's always a good idea to check back in with the problem to be sure that you've addressed all of the issues. I asked you to determine x, $m\angle P$ and $m\angle Q$. You have only found x. You still need to find $m\angle P$ and $m\angle Q$:

$$m\angle P = \frac{x}{2} = \frac{36°}{2} = 18°$$

$$m\angle Q = 2x = 2(36°) = 72°$$

Now your work is done. You have found that x=36°, m∠P = 18°, and m∠Q = 72°.

Eureka!

Review your problem-solving technique:

1. Write down the given information.
2. Draw a picture to visualize the situation.
3. Generate equations based on the given information. Typically, information regarding angle relationships or classifications, like complementary, supplementary, right, straight, and so on can be turned into an equation using the definitions of these terms.
4. Combine the equations and solve for the unknowns.
5. Check back in with the problem statement to be sure that you've addressed all of the questions completely.

A First Look at Proving Angle Congruence

You've spent some time characterizing angles. You've gone in search of relationships between pairs of angles. You've even calculated some angle measurements using your algebra skills. Now it's time to throw another angle into the mix. Although two's company and three's a crowd when it comes to groups of people, angles are a bit more easygoing. In fact, it's kind of fun solving problems with three angles.

In the movie *Dick Tracy*, Dick Tracy had a difficult puzzle to sort out. Was the enemy of his enemy his enemy? Or was the enemy of his enemy his friend? Well, angles have a similar type of problem. Is the supplement of the supplement of an angle a supplement, or is the supplement of the supplement of an angle a complement? There are other convoluted relationships between angles that need sorting out. Sorting out a few of these relationships will help you organize your thoughts and give you some insight into the games that angles play.

> **Example 2:** Suppose ∠A, ∠B, and ∠C have the following relationships: ∠A and ∠B are supplementary, and ∠A and ∠C are supplementary. What can be said about the relationship between ∠B and ∠C?

> **Solution:** The first thing I will do is write down the given information:

> ∠A and ∠B are supplementary, and ∠A and ∠C are supplementary.

> Because ∠A and ∠B are supplementary angles, their measurements add up to 180°, which gives a straight line. The same thing applies to ∠A and ∠C.

The exact size of $\angle A$ is not important to this problem. It doesn't matter whether $\angle A$ is acute, right, or obtuse. $\angle A$ cannot be straight, because straight angles don't have supplements. It is important to note that $\angle A$ is the same size throughout the problem. After a problem begins, angles will not change size. Angles are not allowed to diet or snack while the solution is underway. That goes for all of your other geometric figures as well.

The next step is to draw a picture to visualize the situation. Because I am given that $\angle A$ and $\angle B$ are supplementary, and that $\angle A$ and $\angle C$ are supplementary, I might as well draw both pictures at the same time. Take a look at Figure 4.8. Because of the picture, I can see the relationship between $\angle B$ and $\angle C$: They are congruent. Now, all I have to do is defend my answer.

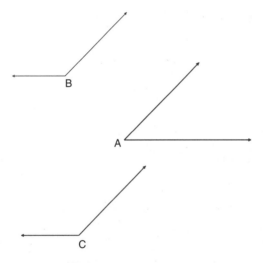

Figure 4.8

$\angle A$ and $\angle B$ are supplementary, as are $\angle A$ and $\angle C$.

The next step is to generate equations based on the given information. Because $\angle A$ and $\angle B$ are supplementary,

$m\angle A + m\angle B = 180°$.

Also, because $\angle A$ and $\angle C$ are supplementary,

$m\angle A + m\angle C = 180°$.

Combining these two equations, I know that

$m\angle A + m\angle B = m\angle A + m\angle C$.

Subtracting $m\angle A$ from both sides of the equation shows that

$m\angle B = m\angle C$,

which means that $\angle B \cong \angle C$.

Finally, let me check back in with the problem statement. What was it that I wanted to establish? I wanted to show that ∠B and ∠C are congruent.

Without the help of my picture, I would not have had any idea about how ∠B and ∠C relate. After I had a picture, I could clearly see the relationship between the two angles. All that was left was to use the definition of supplementary angles to convince the world that the two angles were congruent.

Put Me in, Coach!

Here's your chance to shine. Remember that I am with you in spirit and have provided the answers to these questions in Appendix A.

1. Suppose two angles, ∠P and ∠Q are supplementary, and that $m\angle P = \frac{x}{3}$ and $m\angle Q = 3x$. Find x, m∠P and m∠Q.

2. Suppose ∠A, ∠B, and ∠C have the following relationships: ∠A and ∠B are complementary, and ∠A and ∠C are complementary. What can we say about the relationship between ∠B and ∠C?

The Least You Need to Know

- ◆ Angles can be classified as acute, right, obtuse, or straight.

- ◆ Congruent angles have the same measurement.

- ◆ Complementary angles have measurements that add up to 90°. Supplementary angles have measurements that add up to 180°.

Lines and the Angles They Form

In This Chapter

♦ How lines interact with other lines

♦ Perpendicular and parallel lines

♦ Euclid's 5th postulate

♦ Lots of angle relationships

Lines are one of the fundamental building blocks in geometry. Even though a line is, as previously mentioned, one of three undefined terms, it is probably the most common shape around us. We drive by them, build by them, and stand in them. Anyone who has been to an amusement park can verify that lines seem to go on forever.

When you have a line, there's no limit to what you can make. You can cut it in two and make a ray. You can cut off the ends and make a line segment. From the pieces of a line you have the building blocks to make angles, triangles, and loads of other geometric shapes. You can make practically anything you want!

Fortunately, lines are very well behaved. They don't get bent out of shape or twist the rules around to their advantage. They stay the course, and never change direction. Because of this, it is possible to establish some very nice properties about lines.

Linear Interactions

Lines are so well behaved that whenever you have two lines in the same picture, only one of three things can happen:

- **The two lines occupy the same space.** One line lies directly on top of the other one. Lines cannot bend, and they are both equally straight. The two lines coincide at every point. Because lines have no thickness, you can't even tell that you have two lines. There's no need to worry about the line on the bottom, though. Lines are not heavy (they are made up of points which have no size); the line on the bottom will not be crushed.

- **The two lines meet and intersect.** This interaction is limited: They can meet each other only once. They only have one shot at getting to know each other, and they can only meet at one point. If they decide that they like each other and want to get to know each other better … well, it's against the rules. And lines never bend the rules.

- **The two lines lie next to each other.** These are separated by a constant distance all down the line. In this case, you have two lines that pass through the night, never interacting with each other. They are living so-called "parallel lives," and they will never get to know each other. Their paths will never cross.

Eureka!

Given two lines, only one of three things can happen (1) The two lines are indistinguishable; (2) the two lines intersect at a single point; or (3) the two lines never intersect.

The first option is a bit boring, because you can't be sure that you actually have two lines. The two lines are essentially one and the same. The second and third options are more interesting, and I will spend this chapter exploring how two distinct lines interact.

Intersecting Lines and Vertical Angles

If two straight lines (a.k.a. straight angles) happen to intersect, as in Figure 5.1, four new angles are formed. In this situation, there are more angle relationships than subplots in a soap opera! For example, ∠ABE and ∠ABD share a common vertex (Point B) and a common side ($\overrightarrow{BA}$), but have no common interior points, so they are adjacent angles. But what about ∠ABE and ∠DBC? Those two angles are also related. Although they also share a common vertex (Point B), they have no common side or interior points. But because both ∠DBE and ∠ABC are straight angles, ∠ABE and ∠DBC have a special relationship: When two straight lines intersect, the pairs of nonadjacent angles formed are each known as *vertical angles*. This term might be misleading, because the word "vertical" usually brings up images of things moving up and down. Don't be surprised if you identify vertical angles situated horizontally on the page: ∠ABD and ∠CBE are also vertical angles.

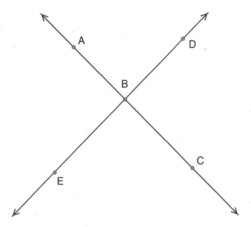

Figure 5.1

$\overleftrightarrow{AC}$ and $\overleftrightarrow{DE}$ *intersect to form four pairs of adjacent angles and two pairs of vertical angles.*

Together ∠DBC and ∠CBE form the straight angle ∠DBE. From the Angle Addition Postulate, you know that m∠DBC + m∠CBE = m∠DBE. Because ∠DBE is straight, m∠DBE = 180°. Thus, m∠DBC + m∠CBE = 180°. So not only are ∠DBC and ∠CBE adjacent angles, they are also supplementary angles.

But wait! There's more! If you act now, you can show that ∠CBE and ∠ABE are also supplementary angles using a similar argument. Remember my observation at the end of Chapter 4: If ∠A and ∠B are supplementary, and ∠A and ∠C are supplementary, then ∠B and ∠C are congruent. You can apply this to the current situation: ∠DBC and ∠CBE are supplementary angles, as are ∠CBE and ∠ABE. That means that ∠DBC and ∠ABE are congruent! So ∠DBC and ∠ABE, which are defined to be vertical angles, are congruent. That's an important idea.

Solid Facts

Vertical angles: a pair of nonadjacent angles formed by the intersection of two straight lines.

Eureka!

When two lines intersect, the vertical angles formed are congruent. This will be used many times to prove angle relationships.

Perpendicular Lines

As I have already mentioned, when two lines intersect, four angles are formed. All four angles cannot be obtuse. If they were, we would have obtuse angles supplementing each other. But an obtuse angle can only be supplemented by an acute angle! So, if you know that one of the angles formed is obtuse, then you actually know that two obtuse angles and two acute angles are formed. The two obtuse angles (which are vertical angles) are congruent, as are the two acute angles. But the angles formed do not need to be obtuse or acute. They could also be just right.

Take a look at Figure 5.2. It shows what happens when two lines intersect to form right angles. Because right angles supplement themselves, and vertical angles are congruent, we can see that if one of the angles formed is right, then all of the angles formed are right. It's a geometrical version of the four musketeers: If one is right, they all are right! This situation deserves a special name and definition. There are actually several ways you can characterize this situation, depending on what you want to emphasize.

Figure 5.2

$\overleftrightarrow{AB}$ *and* $\overleftrightarrow{CD}$ *intersect to form right angles.*

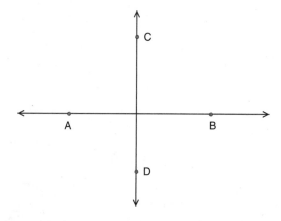

I will show you two different descriptions of this situation and point out why they are equivalent. I will pick one of these definitions and use it throughout this book. There are advantages and disadvantages to each definition. You might not agree with my choice, but when *you* write *your* own geometry book, you can define it any way you want!

One possible name/definition for this situation is as follows: When two lines intersect to form one right angle, the two lines are *perpendicular*.

Another possible definition goes like this: When two lines intersect to form congruent adjacent angles, the two lines are *perpendicular*.

Tangent Line

Mathematicians are constantly developing modern mathematics. New definitions are being created even as I write. Depending on the situation, there are times when the same object is defined differently by different mathematicians. When this happens, someone will have to prove that the various definitions are equivalent. Keep in mind that you can only prove theorems using official definitions, postulates, and previously established theorems.

Let's explore why these two definitions mean the same thing. In the first case, you are forming four right angles. If all four angles are right angles, then all of the angles, and in particular the adjacent angles, are congruent to each other. The first definition is just a simple way to say what the second definition says. In the second definition, the focus is on forming congruent adjacent angles. "Congruent adjacent angles" sounds more intimidating and less intuitive. You need to pick apart the geometric gobbledygook and see what's going on. A picture would be helpful right about now, so turn your eyes to Figure 5.3. In the figure, $\angle 1$ and $\angle 2$ are adjacent angles, so $\angle 1 \cong \angle 2$ by definition.

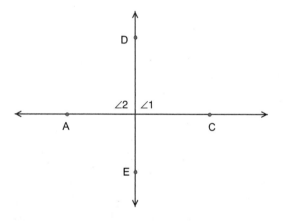

Figure 5.3

$\overleftrightarrow{AC}$ *and* $\overleftrightarrow{DE}$ *intersect to form congruent adjacent angles.*

Also, as discussed before, $\angle 1$ and $\angle 2$ are supplementary angles, so

$$m\angle 1 + m\angle 2 = 180°.$$

Combining these two equations (using the substitution and division properties of equality) and simplifying, you see that

$$2(m\angle 2) = 180°$$

$$m\angle 2 = 90°.$$

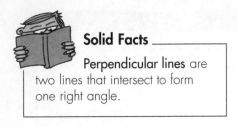

Thus $m\angle 1 = m\angle 2 = 90°$, and both $\angle 1$ and $\angle 2$ are right angles. Applying the relationship between vertical angles, you see that all four angles formed must be right angles. So both definitions mean the same thing, though one definition is more intuitive than the other. Personally, I prefer the first definition, and that's the one I will use throughout the book.

I will use the symbol $\perp$ to indicate a relation "is perpendicular to." If two lines, l and m, are perpendicular to each other, I will write $l \perp m$. Let's see if $\perp$ belongs to the elite class of equivalence relations. Does it have a reflexive property? Is a line perpendicular to itself? No. What about the symmetric property? If $l \perp m$, is $m \perp l$? Yes. Does it have the transitive property? If $l \perp m$, and $m \perp n$, is $l \perp n$? Figure 5.4 shows this situation. Line l cannot be perpendicular to n. So $\perp$ does not have the transitive property. You knew right away that $\perp$ did not belong to the elite class because it failed to have the reflexive property. The only special property that $\perp$ has is the symmetric property.

Figure 5.4

$l\perp m$ *and* $m\perp n$.

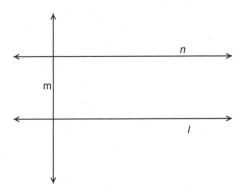

There are some interesting facts about perpendicular lines that you will be able to prove as you progress through this book. One fact in particular has to do with constructing perpendicular lines, and is illustrated in Figure 5.5. Given a line and a point not on that line, there is a unique line passing through the point that is perpendicular to the given line. You will learn how to construct this perpendicular line in Chapter 22, and you will prove this statement in Chapter 10.

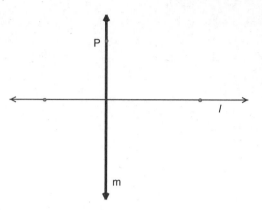

Figure 5.5

Given a line l *and a point* P *not on* l, *there is a unique line* m *passing through* P *perpendicular to* l.

Parallel Lines: Euclid's 5th

In geometry, if two lines are meant for each other, they will intersect. There are some lines that, despite living in the same plane, are destined never to meet: These lines are called *parallel lines*. Parallel lines never intersect. These lines go on forever in two directions, completely unaware that the other one exists.

It takes more than nonintersection to make two lines parallel. The lines must lie in the same plane. Figure 5.6 shows two lines that cannot be classified as being parallel, even though they do not intersect. The reason they are not parallel is that they are not traveling on the same plane.

Solid Facts

Parallel lines are lines that lie in the same plane and do not intersect.

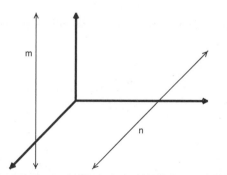

Figure 5.6

Lines m *and* n *do not intersect, yet they are not parallel.*

Euclid's 5th postulate is probably his most controversial. In it, he claims that if you are given a line and a point not on that line, there is a unique line that passes through the point and is parallel to the given line. Many mathematicians (including Euclid himself) believed that this statement could be proven using the other postulates and theorems. None have been successful in doing so. Euclid stated it as a postulate as a last resort. He needed this fact in order to prove theorems about triangles and other geometric shapes. In Chapter 24, I will explore what geometry looks like without this postulate. Don't read ahead! You'll get there soon enough.

> **Postulate 5.2:** The Parallel Postulate. Through a point not on a line, exactly one line is parallel to the given line.

> ### Tangent Line
>
> Neutral Geometry refers to the collection of definitions, postulates, and theorems that can be proven using any of Euclid's postulates *except* his Parallel Postulate. Euclidean Geometry includes Neutral Geometry and any theorems that are proven using the Parallel Postulate.

Euclid's Parallel Postulate looks very similar to the statement about perpendicular lines. In both cases, you are given a line and a point not on the line. In the case of perpendicular lines, you claim that there is a unique line that is *perpendicular* to the given line and passes through the given point. In the Parallel Postulate, you claim that there is a unique line that is *parallel* to the given line and passes through the given point. It doesn't seem right that one statement is a theorem (and can be proven) and the other has to be a postulate (something we believe with our whole heart, but are unable to prove). But that's the way it is.

You've met more new relations in the first five chapters of this book than you would at a family reunion. But there's one more relation that you have to meet: "is parallel to." I will use the symbol $\parallel$ to represent this parallel relation. If two lines, l and m are parallel to each other, I will write l $\parallel$ m. If two lines l and m are not parallel to each other, I will write l $\nparallel$ m.

Does $\parallel$ belong to the elite class? Is a line parallel to itself? Recall that two lines are parallel if they lie in the same plane and do not intersect. By this definition, a line is not parallel to itself. So this relation does not have the reflexive property. What about the symmetric property? If l is parallel to m, is m parallel to l? Yes. Does it have the transitive property? If l is parallel to m, and m is parallel to n, is l parallel to n? Most of the time, in Euclidean geometry, this is true, but it is not always true. For example, if l $\parallel$ m and m $\parallel$ l, then by the transitive property, l $\parallel$ l. But $\parallel$ does not have the reflexive property, so we have a problem. In order for a relation to have the transitive property, it has to always have the transitive property. It's not okay if the relation has the property most of the time. It's an all or nothing kind of deal. So $\parallel$ does not have the transitive property. So $\parallel$ is not an equivalence relation, since it only has the symmetric property.

Transversals and the New Angle Pairs

Imagine that you have two lines, l and m. Just for fun, let's throw in another line, t, which intersects both l and m. I've drawn a picture to help you visualize this (Figure 5.7). The line t is given a special name. Because it cuts across two lines, it is called a *transversal line*.

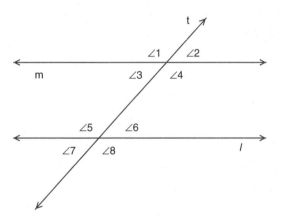

Figure 5.7

Two lines l *and* m, *and a line* t *which intersects both* l *and* m.

There are quite a few angle pairs formed in this situation, and the relationships between all of the different angles need to be sorted out. Because I want to explore how all of the angles relate to each other, I have to come up with some new names. Hang on to your hat, because there's lots of terminology coming your way. As I describe each term, match things up in Figure 5.7.

Solid Facts

A **transversal line** is a line that intersects two (or more) other lines at distinct points.

Angles that are formed between l and m are *interior angles*; those outside l and m are *exterior angles*. In Figure 5.7, ∠3 is an interior angle and ∠1 is an exterior angle. Try to find the other three interior and exterior angles.

The angles that lie in the same position relative to their respective line are called *corresponding angles* for these lines. Corresponding angles must be given as a pair of angles. In Figure 5.7, ∠1 and ∠5 are corresponding angles. Try to find the other three pairs of corresponding angles.

Two interior angles that have different vertices and lie on opposite sides of the transversal are called *alternate interior angles*. In Figure 5.7, ∠3 and ∠6 are alternate interior angles. There's one more pair of alternate interior angles for you to find.

Two exterior angles that have different vertices and lie on opposite sides of the transversal are *alternate exterior angles*. There are two pairs of alternate exterior angles. One pair consists of ∠1 and ∠8. Two interior angles that lie on the same side of the transversal are called *same-side interior angles*. In Figure 5.7, ∠3 and ∠5 are same-side interior angles. Two exterior angles that lie on the same side of the transversal are *called same-side exterior angles*. In Figure 5.7, ∠1 and ∠7 are same-side exterior angles. Whew! That's a lot of angles! Let me remind you that you also have the vertical angles formed between each line and the transversal. You will have your work cut out for you when it comes to exploring how these angles are related and how these angle relationships can be used to characterize lines.

> **Solid Facts** _____
>
> When two lines are cut by a transversal, the following types of angles/angle pairs are formed:
>
> **Interior angles:** Angles that are formed between the two lines.
>
> **Exterior angles:** Angles that are formed outside of the two lines.
>
> **Corresponding angles:** Two angles that lie in the same position relative to their respective line.
>
> **Alternate interior angles:** Two interior angles that have different vertices and lie on opposite sides of the transversal.
>
> **Alternate exterior angles:** Two exterior angles that have different vertices and lie on opposite sides of the transversal.
>
> **Same-side interior angles:** Two interior angles that lie on the same side of the transversal.
>
> **Same-side exterior angles:** Two exterior angles that lie of the same side of the transversal.

The Least You Need to Know

- Given two lines in a plane, they either coincide, intersect at a point, or are parallel.

- When two lines intersect, the vertical angles formed are congruent.

- Two lines are perpendicular if they intersect to form a right angle.

- Euclid's 5th Postulate: given a line and a point not on the line, there is a unique line that passes through the point and is parallel to the given line.

- The relation ⊥ does not have the reflexive or transitive properties, so it is not an equivalence relation.

- The relation ‖ does not have the reflexive or transitive properties, so it is not an equivalence relation.

Chapter 6

A Polygon Is a Many-Sided Thing!

In This Chapter

- ◆ A little bit of Greek
- ◆ What makes a polygon tick
- ◆ Name your polygon
- ◆ Play around with some formulas

Although there are many sides to a polygon, there is not much of depth. Not because there isn't much to learn about polygons, but because they have no thickness.

First there is the practical side of polygons. Polygons play a direct role in automobile safety, for example. Before you embark on a long journey, it's a good idea to check the condition of your vehicle. You wouldn't want something severe to happen, like the engine to fall out. An engine is attached to the frame of the car by several bolts. The heads of these bolts are in the shape of a polygon (usually a hexagon: a six-sided figure). At least one end of the wrench that you use to tighten the bolts will also be in the shape of a hexagon. If your automobile mechanic ever tries to tighten hexagonal bolts with a triangular wrench, you should get a new mechanic!

There is also the mathematical side of polygons. You will want to classify them and get to know their properties. In this chapter, you'll learn how they are made and how to calculate their angle size. You'll also learn what it means to be regular, and you'll make use of the algebra skills you dusted off earlier.

Coming to Terms with the Terminology

You can make polygons by putting line segments together. Each line segment is called a *side* of the polygon. Each endpoint where the sides meet is called a *vertex* of the polygon. The polygon in Figure 6.1 has four vertices: A, B, C, and D.

Figure 6.1

A polygon and its diagonals.

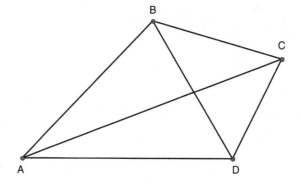

When you move into a new neighborhood, one of the first things you do is get to know your neighbors. Vertices, sides, and angles feel the same way. When they move into a polygon, they want to know what's next to them. If two vertices of a polygon are connected by a side, then they are called *consecutive vertices*. If two sides share a common vertex, then they are called *consecutive sides*. Two angles that share a common side are called *consecutive angles*. Whenever you see the word consecutive, think neighborly thoughts.

A *diagonal* of a polygon is a line segment connecting any two nonconsecutive vertices of the polygon. If the polygon doesn't have too many sides, we can list all of the diagonals by name. The polygon in Figure 6.1 has two diagonals: $\overline{BD}$ and $\overline{AC}$.

> **Tangent Line**
>
> As you become more familiar with polygons, you will notice that a polygon has the same number of sides as it does vertices. At every vertex, there's an angle. Therefore, the number of interior angles of a polygon is equal to the number of vertices of a polygon, which is also equal to the number of sides of a polygon.

> **Solid Facts**
>
> A **side** of a polygon is a line segment used to construct the polygon.
>
> A **vertex** of a polygon is a point of intersection of two sides of the polygon.
>
> **Consecutive vertices** of a polygon are two vertices that share a common side.
>
> **Consecutive sides** of a polygon are two sides that share a common vertex.
>
> **Consecutive angles** of a polygon are two angles that share a common side.
>
> A **diagonal** of a polygon is a line segment connecting any two nonconsecutive vertices of the polygon.

As the number of sides of a polygon increases, it becomes more difficult to count all of the diagonals. Many years ago, some patient mathematicians drew a bunch of polygons and took the time to list (and count) all the diagonals. They made their lists and checked them twice, and eventually derived a formula to calculate the number of diagonals that an *n*-sided polygon has. The total number of diagonals, D, in a polygon with n sides is given by:

$$D = \frac{n(n-3)}{2}$$

Try it out. In Figure 6.1, I showed you a polygon with four sides, and listed two diagonals. Use your formula and see how many diagonals the formula says you should have. If n = 4 then

$$D = \frac{4(4-3)}{2} = \frac{4 \times 1}{2} = 2$$

This formula seems to work. Of course, just checking it with one example isn't enough to convince any hard-core mathematician. Let me assure you that this formula has been proven to work with any polygon you will encounter.

> **Eureka!**
>
> The total number of diagonals, D, in a polygon with n sides is given by:
> $$D = \frac{n(n-3)}{2}.$$

There are two kinds of polygons in the world: convex and concave. The difference between them has to do with the way the vertices are arranged. A *convex polygon* is a polygon in which any segment connecting any two vertices lies inside of the polygon. A *concave polygon* is a polygon in which at least one segment connecting two vertices lies outside the polygon.

> **Solid Facts**
>
> A **convex polygon** is a polygon in which all diagonals lie inside of the polygon.
>
> A **concave polygon** is a polygon in which at least one diagonal is outside the polygon.

Figure 6.2 shows some convex and concave polygons. The idea is that if all diagonals of a polygon stay inside the polygon, the polygon is convex. If even just one diagonal strays outside the polygon, the polygon is concave.

Figure 6.2

Some convex polygons (top) and concave polygons (bottom).

In order to build a polygon you have to put the segments together carefully; you can't just throw a bunch of segments into a pile and call it a polygon. There are rules for making polygons that you have to follow.

1. Segments are only allowed to touch each other at their endpoints. They aren't allowed to touch each other in the middle, or cross each other. And segments definitely aren't allowed to sit on top of each other.

2. Each endpoint of a segment must be connected to another segment. Because a segment has two endpoints, each segment must come in contact with two other segments. The result of this second rule is that our segments will join to form closed geometric shapes.

3. All of the segments must lie in the same plane. You don't want your polygons sticking any of their segments out of their plane. That's how they can get broken off.

4. The final criteria for the polygons discussed in this book will be their shape. As a card-carrying member of the Mathematicians Union of the Geometric World, my union will only allow me to work with convex polygons. As my apprentice, you are subject to the same restrictions. Unless otherwise stated, all polygons in this book are convex.

Solid Facts

A **polygon** is a closed plane figure whose sides are line segments that intersect only at the endpoints.

Now that you have an idea of how a polygon is made, let's define one. A *polygon* is a closed plane figure whose sides are line segments that intersect only at the endpoints.

Naming Conventions and Classifications

Now that you know what a polygon is, let me introduce you to the many types of polygons that are out there. It will also be to your advantage if you learn how to call them by name. They don't usually answer, but this way you will be able to talk about them even when they aren't around. I will focus on the different *types* of polygons first.

I would love to be able to tell you that classifying polygons involves rules that are never broken. But I would be lying. Although there are conventions to follow, the exceptions come first. After those exceptions, however, things settle down and the rules are obeyed. Categorizing polygons with five sides or more requires familiarity with Greek prefixes. That won't be a problem though, because these prefixes have become ingrained in the English language. For example, you know how many wheels a bicycle or a tricycle has without a second thought.

There is no such thing as a two-sided polygon. Every polygon has at least three sides. The three-sided polygons are probably the most famous of them all, and you will study them in detail in Part 3. They are officially called *triangles*. The prefix *tri-* means three, so the name triangle indicates that the polygon has three angles. The triangle is the only polygon whose name explicitly states the number of angles that it has.

A four-sided polygon is called a *quadrilateral* (it's described by the number of sides that it has). That's it as far as the exceptions go. Use the following guidelines from now on. A polygon's type depends on the number of sides, and every polygon type ends in *–gon*. The following table lists the number of sides, the prefix, and the official polygon classification for the first twelve polygons and the general case.

> **Tangent Line**
>
> Different traffic signs have different geometric shapes. Speed limit signs are quadrilaterals, a yield sign is in the shape of a triangle, and a stop sign is an octagon.

Naming Polygons

Sides	Prefix	Name	Sides	Prefix	Name
3	Tri	Triangle	9	Nona	Nonagon
4	Quad	Quadrilateral	10	Deca	Decagon
5	Penta	Pentagon	11	Undeca	Undecagon
6	Hexa	Hexagon	12	Dodeca	Dodecagon
7	Hepta	Heptagon			
8	Octa	Octagon	n		n-gon

Whenever you want to refer to a polygon, you first need to state the type of polygon you are working with. Then you need to list, in succession, the capital letters representing consecutive vertices. For example, the polygon in Figure 6.3 would be referred to as *quadrilateral ABCD*. You could also refer to it as *quadrilateral BCDA*. You can list the vertices as they occur clockwise or counter-clockwise, and you can start anywhere you like. Just make sure that you go completely around and list each vertex, in consecutive order, only once.

Figure 6.3

The quadrilateral ABCD.

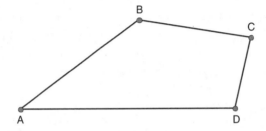

The Interior Angles

So you've got a polygon. Now what? Well, you can discover some unusual properties of polygons. You've already learned that the number of diagonals of a convex polygon depends only on the number of vertices that the polygon has, and not how they are arranged. Speaking of vertices, wherever there's a vertex, there's an angle. Polygons have some special angle relationships worth mentioning.

If you're going to study angle relationships, it's best to start with the polygon that has "angle" as its last name—the triangle. If you measure all of the interior angles of a triangle and add them together, the sum will always be 180°. It doesn't matter how big or small the triangle is. The sum of the interior angles will always be 180°. To prove this you must use Euclid's 5th Postulate, which means that this result only holds in Euclidean Geometry. Because the sum of the measures of the interior angles of a triangle is 180°, you can establish a method for finding the sum of the interior angles of any type of polygon. This method involves using some of the diagonals of the polygon to divide it up into several triangles and then applying the Angle Addition Postulate. For example, you can use one of the diagonals of a quadrilateral to break it up into two triangles, as shown in Figure 6.4. Using the Angle Addition Postulate, you can show that the sum of the interior angles of a quadrilateral is always 360°.

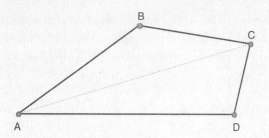

Figure 6.4

The diagonal of a quadrilateral divides the quadrilateral into two triangles.

If you can break a quadrilateral into two triangles, you can certainly break an n-sided polygon into a bunch of triangles and use them to determine the sum of the polygon's interior angles. In fact, if you did this for a pentagon and a hexagon, it wouldn't take long for us to observe the pattern. The formula for the sum of the interior angles of a polygon, S, as a function of the number of sides of the polygon, n, is

$$S = (n - 2) \times 180°$$

The factor (n – 2) tells you how many triangles you can make in your polygon. When n = 4 (a quadrilateral), you can divide it up into 2 triangles. When n = 5 (a pentagon), you can divide it up into 3 triangles, as is shown in Figure 6.5.

Eureka!

The sum of the interior angles of a polygon having n sides is given by $S = (n - 2) \times 180°$.

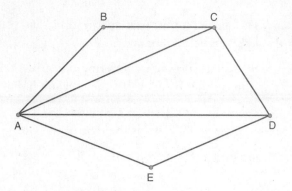

Figure 6.5

The pentagon ABCDE with diagonals from one vertex illustrates how to break up a pentagon into triangles in order to find the sum of the interior angles.

Examine Figure 6.5 in more detail. The diagonals $\overline{AC}$ and $\overline{AD}$ break the pentagon into three triangles: $\triangle AED$, $\triangle ADC$, and $\triangle ABC$. The sum of the measures of the interior angles of the pentagon is given by:

$$m\angle A + m\angle B + m\angle C + m\angle D + m\angle E$$

Because the sum of the measures of the interior angles of a triangle is 180°, and we have three triangles, you can add up the interior angles of each triangle together. Then you can connect the interior angles of the triangles with the interior angles of the pentagon:

$$m\angle 1 + m\angle 2 + m\angle 3 = 180°$$

$$m\angle 4 + m\angle 5 + m\angle 6 = 180°$$

$$m\angle 7 + m\angle 8 + m\angle 9 = 180°$$

If you add these three equations together, and group things together in such a way as to make use of the Angle Addition Postulate, you obtain the following result:

$$m\angle 1 + m\angle 2 + m\angle 3 + m\angle 4 + m\angle 5 + m\angle 6 + m\angle 7 + m\angle 8 + m\angle 9 = 3 \times 180°$$

$$m\angle 1 + m\angle 6 + m\angle 9 + m\angle 2 + m\angle 3 + m\angle 4 + m\angle 5 + m\angle 7 + m\angle 8 = 540°$$

$$m\angle A + m\angle B + m\angle C + m\angle D + m\angle E = 540°$$

I'll work some examples to show you what this formula can do.

Example 1: Find the sum of the measures of the interior angles of a polygon with 5 sides.

Solution: Using your equation with n = 5, you have

$$S = (5 - 2) \times 180° = 3 \times 180° = 540°$$

This is the same result you obtained using "triangularization," but this time you didn't have to suffer through as much algebra.

If you are given the sum of the measures of the interior angles of a polygon, you can use your equation to find the number of sides that the polygon has.

Example 2: Find the number of sides that a polygon has if the sum of the measures of its interior angles is 900°. What is the type of this polygon?

Solution: Substituting into our equation and simplifying, we see that:

$$900° = (n - 2) \times 180°$$

$$900° = 180° \times n - 360°$$

$$180° \times n = 1260°$$

$$n = \frac{1260°}{180°} = 7$$

So the polygon has 7 sides, making it a heptagon.

Regular Polygons

There's nothing run of the mill about regular polygons. They are actually quite special. With generic polygons the segments can be any length they want to be, as long as the final result is convex. Some really long sides can be mixed with some really short sides to make a single polygon, and although it might look funny, it wouldn't violate any of our rules.

Even though polygons aren't concerned with fashion trends, you might have some preconceived ideas about what makes a polygon look good. As you construct polygons, you can be particular about a polygon's proportions. When you start putting restrictions on polygons, they get closer to being "regular." For example, if you construct a polygon using only congruent segments for the sides, then you are constructing an equilateral polygon. If you make a polygon using only congruent interior angles, then what you have is an equiangular polygon.

But being equilateral or equiangular is not enough to earn the title "regular polygon." A *regular polygon* is a polygon that is *both* equiangular and equilateral. Figure 6.6 shows polygons that are, respectively, (a) equilateral, (b) equiangular, and (c) regular.

Solid Facts

An **equilateral polygon** is a polygon with all sides congruent.

An **equiangular polygon** is a polygon with all angles congruent.

A **regular polygon** is a polygon that is both equilateral and equiangular.

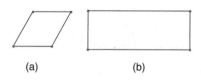

(a)   (b)   (c)

Figure 6.6

Polygons that are (a) equilateral, (b) equiangular, and (c) regular.

If you have an equiangular polygon or a regular polygon with n sides, you can develop a formula to calculate the size of each angle. All of the angles of both regular and equiangular polygons are congruent. Now, an n-sided polygon has n interior angles, and each has measure I. So the sum of the interior angles of our regular or equiangular polygon must be n × I. You already have a formula that you can use to calculate the sum of the interior angles: S = (n – 2) × 180°. Substituting in for the sum of the interior angles and simplifying yields:

$$n \times I = (n - 2) \times 180°$$

$$I = \frac{(n - 2) \times 180°}{n}$$

Eureka!

Given an n-sided equiangular or regular polygon, the measure, I, of each interior angle is given by the formula

$$I = \frac{(n - 2) \times 180°}{n}$$

This formula can be used to compute either the size of the interior angles of an n-sided regular or equiangular polygon. You can also use this formula to compute the number of sides of a regular or equiangular polygon if you are given the size of each interior angle. I will work examples of each type of problem.

Example 3: Find the measure of each interior angle of an equiangular octagonal ceramic floor tile, as shown in Figure 6.7.

Figure 6.7

An equiangular octagonal ceramic floor tile used in Example 3.

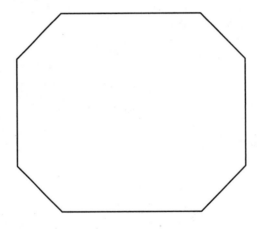

Solution: For an octagon, n = 8. Substituting the value of n into our equation yields:

$$I = \frac{(8 - 2) \times 180°}{8} = \frac{6 \times 180°}{8} = 135°$$

So each interior angle measures 135°.

Example 4: Each interior angle of a certain regular polygon has a measure of 144°. Find the number of sides it has, and identify the type of polygon it is.

Solution: Suppose that the polygon has n sides. You are given that the size of the interior angles is 144°. Substituting this into our formula, and simplifying, you have:

$$144 = \frac{(n-2) \times 180°}{n}$$

$$144° \times n = (n-2) \times 180°$$

$$144° \times n = 180° \times n - 360°$$

$$36° \times n = 360°$$

$$n = \frac{360°}{36°} = 10$$

So your regular polygon has 10 sides, and it is a regular decagon.

Put Me in, Coach!

Here's your chance to shine. Remember that I am with you in spirit and have provided the answers to these questions in Appendix A.

1. How many diagonals does a polygon with six sides have?

2. Find the sum of the measures of the interior angles of a hexagon using your new formula.

3. Find the number of sides in a polygon whose sum of interior angles is 2340°.

4. Find the measure of each interior angle of a regular polygon having 6 sides.

5. Find the number of sides that a regular polygon has if the measure of each interior angle is 150°.

The Least You Need to Know

♦ The total number of diagonals of a polygon can be found using the formula $D = \frac{n(n-3)}{2}$.

♦ A polygon is categorized by the number of sides that it has. To name a polygon, use the appropriate Greek prefix in front of *–gon*.

♦ The sum of the interior angles of a polygon can be found using the formula $S = (n-2) \times 180°$.

♦ A regular polygon is a polygon that is both equilateral and equiangular. The size of the interior angles of a regular polygon can be found using the formula $I = \frac{(n-2) \times 180°}{n}$.

Part 2

Introducing Proofs

There's one thing you can say about Euclid and his gang: They were extremely logical. If Euclid were alive today, he would have been made an honorary Vulcan. In geometry, nothing happens by chance, and there are no miracles. There is an explanation for everything that happens. In this part, you will develop your logic and reasoning skills and use them to explain some amazing discoveries.

Though there are no miracles in geometry, there is room for faith. But the goal is to minimize what you take on faith (the postulates) and maximize what you can explain (the theorems).

Chapter 7

Logic: Rules for Arguing

In This Chapter

♦ How to induct and when to deduct

♦ Creating truth tables

♦ Conjunction, disjunction, and implication

♦ Logical equivalence and tautological statements

There are two types of reasoning in mathematics: inductive and deductive. The well-known authors Agatha Christie and Sir Arthur Conan Doyle managed to make logical reasoning fashionable, at least in their version of the criminal underworld. They both made you think that their mysteries could be solved by applying a bit of logic.

Inductive reasoning involves observing patterns and making general statements based on those observations. This type of reasoning was made famous by Agatha Christie's literary character Miss Marple. She would often correctly determine the culprit based on the patterns of behavior of old relatives and friends.

Deductive reasoning, on the other hand, was made famous by Sir Arthur Conan Doyle's character Sherlock Holmes. Sherlock Holmes is famous for drawing conclusions based on statements that are accepted as true. With deductive reasoning, valid arguments are used to show why one thing follows from another.

Inductive Reasoning

Remember those patient mathematicians who drew hundreds of polygons and counted the diagonals? Their hard work paid off, because they eventually discovered a relationship between the number of diagonals of a polygon and the number of sides. Those mathematicians were practicing what we call *inductive reasoning*. Inductive reasoning is the process of observing a situation and making a generalization based on those observations.

Inductive reasoning isn't just for mathematicians, and there's no law that requires us to only use it in a classroom. Inductive reasoning helps with recognizing patterns and making predictions based on past experience. Pattern recognition is used extensively when playing around with numbers. Ancient mathematicians used to study numbers that corresponded to geometric figures. For example, square numbers are numbers that correspond to ... you guessed it: squares! Figure 7.1 shows a few square numbers. Let's see if we can recognize a pattern.

> **Solid Facts**
>
> **Inductive reasoning** is the process of observing a pattern and drawing a general conclusion from those observations.

Figure 7.1

Square numbers.

Example 1: Find a pattern in determining the square numbers, and find a formula to find the *n*th square number.

Solution: As usual, a picture is worth a thousand words. With the help of Figure 7.1, you can easily determine the pattern. The sequence of square numbers is 1, 4, 9, and so on. Remember your multiplication lessons from elementary school? One times one is one (the first square number!), two times two is four (the second square number!), three times three is nine (the third square number!); the fourth square number is sixteen (four times four). The *n*th square number is n times n, or n^2.

Although inductive reasoning can be useful in some situations, it does have its limitations. With this type of reasoning you are drawing conclusions based on only a few observations. If you quit observing too soon, you might miss something important. It's never quite clear how long you must observe before you find the correct pattern.

And there's always the risk that any pattern you find might fail further down the road. There's no guarantee that you'll be able to find a pattern, or even that a pattern exists. You might think that's bad, but it gets even worse. You might have a situation where you find several patterns, and you can't tell the difference between the true pattern and a clever imposter. Even with its limitations, inductive reasoning is used extensively in mathematics.

CAUTION Tangled Knot

Just because you are looking for patterns doesn't mean that you'll be able to find just one. Sometimes you might see two or more patterns, and other times no pattern will appear. For example, the list "2, 4, ..." can be followed with a 6 or with an 8, depending on your interpretation. You might think that the next term is obtained by *adding* 2 to 4 to get 6, or you might believe that the next term is obtained by *doubling* 4 to get 8. When this happens, you will need to examine more terms in the sequence to distinguish between the two possibilities.

Deductive Reasoning: Elementary, My Dear Watson!

Deductive reasoning is the type of reasoning used when you derive a conclusion from other *statements* that you accept as true. The words statements and sentences are usually interchangeable, but in mathematics the two words mean very different things. Sentences are required to have a subject and a predicate (and before you go and check the title of the book again, that's all the English I plan to discuss!). Statements, on the other hand, have very specific requirements. A statement is a group of words or symbols that can be classified collectively as either true or false. For example, the claim "2 + 3 = 5" is a true statement, whereas "A triangle has four sides" is a false statement. But "Help me!" is not a statement at all (though it is a sentence).

Solid Facts

A **statement** is a group of words or symbols that can be classified collectively as true or false.

You were exposed to deductive reasoning in the previous chapters, though you weren't properly introduced. For example, in Chapter 5 you learned that vertical angles are congruent, and you used deductive reasoning to draw that conclusion.

Although examples of deductive reasoning can be found in detective stories and murder mysteries, deduction isn't just for criminal investigators. It can also be used in a variety of everyday situations. For instance, I know that if a police officer catches me speeding I will get a ticket. (You can use inductive reasoning to verify that that statement is true, but don't expect me to help pay the fine!) Using deductive reasoning, I know what to expect if I speed past a patrol car as I hurry to get to where I'm going.

> **Eureka!**
>
> Deductive reasoning and inductive reasoning often go hand in hand. Mathematicians usually use inductive reasoning to observe patterns and make conjectures, and then they use deductive reasoning to determine if the discoveries are logically consistent.

Now that you have some of the pieces to the puzzle, you can put them together and start to see the big picture. The statements mathematicians accept as true are either premises or postulates. Premises and the conclusions that you draw from them are what theorems are all about. The steps that you take to go from the premises to the conclusion constitute your proof, and your proofs are put together using deductive reasoning.

Logical Constructions and Truth Tables

If you want to draw conclusions from a collection of premises, you need to construct a logical argument. A logical argument requires you to string enough true statements that are dependent on each other to get from the assumptions to the conclusion. Because you are dealing with a collection of statements, you'll need some logical tools to help you work with them. The first tool at your disposal is the truth table. A *truth table* is a table that shows all of the relationships between given statements. Truth tables can involve as many statements as you like. I'll keep things simple and work with two statements.

> **Solid Facts**
>
> A **truth table** is a table that provides the truth values of a statement by considering all possible true/false combinations of the statement's components.

The statements themselves can get very wordy. It doesn't matter what the statements actually say, all that's important is whether they are true or false. To keep things simple I'll use letters like P and Q to represent statements.

Negation

If you are in a negative mood and you have a statement hanging around, you might want to negate it as well. If you negate the statement P, you will write it as ~P (and read it "not"). Remember that all statements can be classified as either true or false. If P happens to be true, then ~P will be false. If P is false, then ~P will be true. P and ~P are always opposites of each other. You can build your first truth table that shows the relationship between P and ~P. In truth tables it is standard to write T for true and

F for False. The first column in a truth table is the statement P, the second column is the statement ~P. Now, P has only two options available: It can be either true or false. I'll put each option as a separate row in the following truth table. The first row in the truth table shows that if P is true, then ~P is false. The second row shows that if P is false, then ~P is true.

Truth table for P **and** ~P

P	*~P*
T	F
F	T

Conjunction

Individual statements can be combined to form compound statements. For example, a statement of the form "P and Q" is called the *conjunction* of P and Q. In symbols, the conjunction is written $P \wedge Q$. In order for the conjunction to be true, it is necessary that both P and Q are true. If either statement is false, the conjunction is false.

In order to construct a truth table for the compound statement $P \wedge Q$, we need to explore all possible combinations of truth values for both P and Q. When P is true, Q may be true or false. Similarly, when P is false, Q may be true or false. So the truth table must have four rows, one for each combination of truth values for P and Q.

Solid Facts

The **conjunction** of P and Q is a statement of the form "P and Q."

Tangled Knot

When constructing a truth table, you should always begin the same way: the first two columns should be the truth values for the statements P and Q. Always write the truth values in the same pattern. This will make it easy to compare truth tables.

Truth table for $P \wedge Q$

P	*Q*	$P \wedge Q$
T	T	T
T	F	F
F	T	F
F	F	F

Example 2: Let P be the statement "Fire is hot" and Q be the statement "$3 + 4 < 5$". Is $P \wedge Q$ true or false?

Solution: Because Q is false, $P \wedge Q$ is false. It doesn't matter whether P is true or false. All it takes is one of the statements to be false to make the compound statement false.

Disjunction

A compound statement of the form "P or Q" is called the *disjunction* of P and Q. In symbols, the disjunction is written $P \vee Q$. The only way that $P \vee Q$ is false is if *both* P and Q are false. All it takes for is $P \vee Q$ to be true is for either P or Q to be true. If both P *and* Q happen to be true, that's just icing on the cake. $P \vee Q$ will also be true. To help you understand this, consider the following situation: you can try out for the baseball team if you have a B average in school or you have played on a little league team for at least three years. If you satisfy both requirements, you may still try out for the team.

> **Solid Facts**
>
> The **disjunction** of P and Q is a compound statement of the form "P or Q."

Truth table for $P \vee Q$

P	Q	$P \vee Q$
T	T	T
T	F	T
F	T	T
F	F	F

> **Tangent Line**
>
> The disjunction is often referred to as an *inclusive or*. *Inclusive* means that both options are available simultaneously. Restaurants never use an "inclusive or". When given the choice of soup or salad, try interpreting the "or" as an "inclusive or" and see what you get.

You can create even more complicated compound statements by combining *ands* and *ors*. In some cases, you can use parentheses to clarify the meaning of a compound statement. The parentheses are given priority just as they are in algebraic expressions. To build a truth table for a complicated compound statement, just pick apart the statement piece by piece. Start with P and Q, and then build the statements inside parentheses and work your way out. You can see the logical progression in the columns of the truth table for Example 3.

Example 3: Construct a truth table for the compound statement $P \wedge (\sim P \vee Q)$.

Solution: Begin with your truth values for P and Q, and then start building. First build $\sim P \vee Q$. To do this you'll need a $\sim P$ column. After you build $\sim P \vee Q$ you can combine it with P to finally determine the truth values for the compound statement $P \wedge (\sim P \vee Q)$.

Constructing a truth table for $P \wedge (\sim P \vee Q)$

P	Q	$\sim P$	$\sim P \vee Q$	$P \wedge (\sim P \vee Q)$
T	T	F	T	T
T	F	F	F	F
F	T	T	T	F
F	F	T	T	F

Implication (or Conditional)

The final compound statement to consider is of the form "If P then Q." This statement is called an *implication* or a *conditional* statement. In symbols, you write $P \rightarrow Q$ (and read it "if P then Q"). P is called the hypothesis (the given information) and Q is called the conclusion (it's what you want to prove). The conditional statement makes a promise. The promise is broken if P is true and Q is false. Consider the statement "If you clean your room, I will give you some ice cream." To determine the truth values for $P \rightarrow Q$, it's best to think about things in terms of broken promises. If it happens that you clean your room *and* I give you ice cream, then there's no problem: I kept my promise. The truth value of $P \rightarrow Q$ in this case is true. If you clean your room and I don't give you ice cream, then I broke my promise and you have a legitimate complaint. The truth value of $P \rightarrow Q$ in this case is false. If you didn't clean your room and still I gave you ice cream, then I'm just a nice person, and you would probably not complain. The truth value of $P \rightarrow Q$ would be true. If you didn't clean your room and I didn't give you ice cream, you would still have no complaint, because you didn't do what I asked you to do. The truth value of $P \rightarrow Q$ is true once again. The only time that the truth value of $P \rightarrow Q$ is false is if I break my promise: The statement P is true (the condition is met) and the statement Q is false (no follow-through). The truth table for $P \rightarrow Q$ is shown below.

Truth table for $P \rightarrow Q$

P	Q	$P \rightarrow Q$
T	T	T
T	F	F
F	T	T
F	F	T

Solid Facts

An **implication** or a **conditional statement** is a statement of the form "If P then Q."

Example 4: Construct a truth table for the compound statement $(P \rightarrow Q) \vee P$.

Solution: As usual, we start with columns for P and Q. Then we pick apart the compound statement, starting with the parentheses. We need a column for $P \rightarrow Q$. When that's done, we can determine the truth values for $(P \rightarrow Q) \vee P$.

Truth table for $(P \rightarrow Q) \vee P$

P	Q	$P \rightarrow Q$	$(P \rightarrow Q) \vee P$
T	T	T	T
T	F	F	T
F	T	T	T
F	F	T	T

If you start with the conditional statement $P \rightarrow Q$, it matters which statement is the hypothesis and which is the conclusion. $P \rightarrow Q$ is quite different from $Q \rightarrow P$, as can be seen from the room-cleaning/ice cream promise. $Q \rightarrow P$ is called the *converse* of the statement $P \rightarrow Q$. It's easy to see that the two conditional statements are different by constructing a truth table. If you are in a particularly contrary mood, you might want to negate your statements P and Q. The statement $\sim Q \rightarrow \sim P$ is called the *contrapositive* of the statement $P \rightarrow Q$, and $\sim P \rightarrow \sim Q$ is the *inverse* of the statement $P \rightarrow Q$.

Eureka!

$P \wedge Q$ is true only when both P and Q are true. $P \vee Q$ is false only when both P and Q are false. $P \rightarrow Q$ is false only when P is true and Q is false.

Example 5: If P is the statement "it is raining" and Q is the statement "the sky is cloudy," write the conditional ($P \rightarrow Q$), the converse, the inverse, and the contrapositive compound statements using words, not symbols.

Solution: The conditional statement: "If it is raining, then the sky is cloudy." The converse: "If the sky is cloudy, then it is raining." The inverse: "If it is not raining, then the sky is not cloudy." The contrapositive: "If the sky is not cloudy, then it is not raining."

Solid Facts

The statement $P \rightarrow Q$ is called a **conditional statement**.

The statement $Q \rightarrow P$ is the **converse** of $P \rightarrow Q$.

The statement $\sim P \rightarrow \sim Q$ is the **inverse** of $P \rightarrow Q$.

The statement $\sim Q \rightarrow \sim P$ is the **contrapositive** of $P \rightarrow Q$.

I will explore the relationships between all possible combinations of the conditional statement in the next section.

Logical Equivalence and Tautology

When I first showed you how to build truth tables, I recommended that you always start out the same way: the first two columns are reserved for P and Q, and you fill in the truth values for P and Q using the same pattern. The reason for this is that you will want to compare truth values for different compound statements and search for patterns. For example, the truth values for $P \wedge (\sim P \vee Q)$ in Example 3 look a lot like the truth values for $P \wedge Q$. In situations like this, mathematicians say that $P \wedge (\sim P \vee Q)$ and $P \wedge Q$ are logically equivalent. Two statements are *logically equivalent* if their truth values are the same for all possible true/false combinations of their components. A double arrow is used to show that two statements are logically equivalent.

Example 6: Use a truth table to show that the statements $P \rightarrow Q$ and $\sim Q \rightarrow \sim P$ are logically equivalent.

Solution: You have already constructed the truth table for $P \rightarrow Q$, so the real work is in constructing the truth table for $\sim Q \rightarrow \sim P$. To build the truth table for $\sim Q \rightarrow \sim P$ you first need $\sim Q$ and $\sim P$.

Solid Facts

Two statements are **logically equivalent** if their truth values are the same for all possible true/false combinations of their components.

Truth table for ~ Q → ~ P and P → Q

P	Q	~ Q	~ P	~ Q → ~ P	P → Q
T	T	F	F	T	T
T	F	T	F	F	F
F	T	F	T	T	T
F	F	T	T	T	T

Notice that the truth values for $\sim Q \to \sim P$ and $P \to Q$ match each other, row for row. A conditional statement is logically equivalent to its contrapositive. You would write $(\sim Q \to \sim P) \leftrightarrow (P \to Q)$.

Solid Facts

A **tautology** is a statement that is true for all possible truth values of its components.

Revisit the truth table for $(P \to Q) \vee P$ in Example 4. All of the truth values for $(P \to Q) \vee P$ are true. When that happens, your statement is called a *tautology*. A tautology is a statement that is true for all possible truth values of its components.

The compound statement $P \to Q$ is not a tautology, because there is one truth value that is false. All it takes is one entry to be false to ruin the whole thing.

It might seem strange that an entire chapter of this geometry book is devoted to building truth tables, especially because I haven't mentioned points, lines, or anything remotely resembling something geometric for awhile. There is a method to my madness: The reason for this detour into logic land is that tautologies are used to construct geometric proofs. In order to prove a mathematical statement, you need to use a valid argument. To establish that an argument is valid, you must show that its premises and conclusion form a compound statement that is a tautology. I'll explore the logic behind some common valid arguments in the next chapter.

Put Me in, Coach!

Here's your chance to shine. Remember that I am with you in spirit and have provided the answers to these questions in Appendix A.

1. Consider the rectangular numbers shown in Figure 7.2. Find the sequence of numbers, and find the general pattern.

Figure 7.2

Rectangular numbers

2. Which of the following are statements? For each one that is a statement, is it true or false?

   a. 4 + 3 < 5

   b. Babe Ruth played professional sports.

   c. How are you?

3. Let P be the statement "Fire is hot" and Q be the statement "3 + 4 < 5". Is

   $P \wedge (\sim Q)$ true or false?

4. Construct a truth table for the compound statement $P \wedge (\sim P \vee Q)$.

5. Construct a truth table for the compound statement $\sim P \rightarrow \sim Q$.

6. If P is the statement "the measure of $\angle A$ is less than 90°" and Q is the statement "$\angle A$ is acute," write the conditional ($P \rightarrow Q$), converse, inverse, and contrapositive compound statements using words, not symbols.

The Least You Need to Know

- When you observe patterns and generalize what you discover, you are using inductive reasoning.

- When you draw conclusions from statements that are accepted as true, you are using deductive reasoning.

- $P \wedge Q$ is only true if both P and Q are true; $P \vee Q$ is only false if both P and Q are false; $P \rightarrow Q$ is only false if P is true and Q is false.

- To construct truth tables of compound statements, build compound statement piece by piece, starting with whatever is inside parentheses.

Chapter 8

Taking the Burden out of Proofs

In This Chapter

- ◆ Prove things directly
- ◆ Prove things indirectly
- ◆ Construct truth tables
- ◆ Write a two-column proof

When a kid makes exaggerated claims out on the playground, usually someone will ask the bragger to "prove it." What follows is usually an awkward silence while the bragger thinks about how to justify his boast.

When mathematicians make exaggerated claims in a classroom, usually no one will ask them to "prove it." In fact, most people are happier not knowing why a particular mathematical statement is true. But not mathematicians. It's not enough to brag about the properties of angles, segments, or other geometric shapes. They back up their claims whenever someone dares to doubt them.

When mathematicians are backing up their claims, they don't lead the audience on any wild goose chases. They don't get side tracked either. When mathematicians prove a statement the explanation is short, concise, and to the point. Organization is the key to writing proofs. In this chapter, you will learn some ways to organize your arguments so that everyone will understand them and there will be no room for doubt.

The Law of Detachment

The first type of valid argument I will discus is commonly known as the law of detachment. To illustrate the law of detachment, consider the following example.

> **Example 1**: Assume that the two premises below are true. What conclusions can be drawn?
>
> 1. If it is raining, the baseball game will be cancelled.
>
> 2. It is raining.
>
> **Solution**: If you accept these statements to be true, then you can logically conclude that the baseball game will be cancelled. You can examine this type of argument easier if you switch over to P's and Q's. Let P be the statement "it is raining" and Q be the statement "the baseball game will be cancelled." Then your first statement is $P \rightarrow Q$ and the second statement is P. The conclusion is that the baseball game will be cancelled, which is Q. So you are assuming two things: $P \rightarrow Q$ and P, and claiming that Q follows. The compound statement for assuming $P \rightarrow Q$ and P, and concluding Q can be written $((P \rightarrow Q) \wedge P) \rightarrow Q$. Now that's a complicated compound statement! Let's build a truth table for $((P \rightarrow Q) \wedge P) \rightarrow Q$ and see what you get.

Truth table for $((P \rightarrow Q) \wedge P) \rightarrow Q$

P	Q	$P \rightarrow Q$	$(P \rightarrow Q) \wedge P$	$((P \rightarrow Q) \wedge P) \rightarrow Q$
T	T	T	T	T
T	F	F	T	T
F	T	T	T	T
F	F	T	T	T

Notice that $((P \rightarrow Q) \wedge P) \rightarrow Q$ is a tautology, and hence is a valid argument. So you can logically conclude that the baseball game will be cancelled in this situation.

Let's look at another example.

> **Example 2**: Assume that the two premises below are true. What conclusions can be drawn?
>
>   1. If it is raining, the baseball game will be cancelled.
>
>   2. The baseball game is cancelled.

> **Solution**: It is tempting to conclude that it is raining. But the game could have been cancelled for another reason; maybe one team didn't have enough players. You can examine this scenario using formal logic as well. As you did before, let P be the statement "it is raining" and Q be the statement "the baseball game will be cancelled." Then the first statement is $P \rightarrow Q$ and the second statement is Q. The conclusion is that it is raining, which is just P. Again we are assuming two things: $P \rightarrow Q$ and Q, and claiming that P follows. The compound statement for assuming $P \rightarrow Q$ and Q, and concluding P can be written $((P \rightarrow Q) \wedge Q) \rightarrow P$. Let's build the truth table for this compound statement and see what we get.

Truth table for $((P \rightarrow Q) \wedge Q) \rightarrow P$

P	Q	$P \rightarrow Q$	$(P \rightarrow Q) \wedge Q$	$((P \rightarrow Q) \wedge Q) \rightarrow P$
T	T	T	T	T
T	F	F	F	T
F	T	T	T	F
F	F	T	F	T

Because one of the truth values for $((P \rightarrow Q) \wedge Q) \rightarrow P$ is false, $((P \rightarrow Q) \wedge Q) \rightarrow P$ is not a tautology, and you cannot draw any conclusions about P (whether or not it is raining). It's nice to see that our formal logic confirms what our intuitive logic led us to believe.

The Importance of Being Direct

The next type of argument is the direct proof, which can be interpreted as a transitive property of implication. Consider these two statements:

If it is raining, the street is wet.

If the street is wet, then the street is slippery.

You can eliminate the middleman and draw the conclusion "if it is raining, then the street is slippery" directly. This is the type of reasoning that Sherlock Holmes is famous for. He omits the middle statements and draws what appear to be surprising and clever conclusions as a result.

You can put this into a logical format and analyze it using P's Q's and R's. You haven't analyzed three statements together yet, but it's not much different than analyzing two statements together. You just have a lot more options for the truth value combinations. In fact, when you form a compound statement using three individual statements, your truth table will have 8 rows.

Tangent Line

You can determine how many rows a truth table with four statements would require using inductive reasoning. A truth table involving one statement requires two rows. A truth table involving two statements requires four rows. When a truth table involves three statements, use eight rows. Following the pattern of doubling, a truth table involving four statements would need sixteen rows. In fact, a truth table involving n rows requires 2^n rows.

If P is the statement "it is raining", Q is the statement "the street is wet," and R is the statement "the street is slippery," then your premises are $P \rightarrow Q$ and $Q \rightarrow R$. Your conclusion is $P \rightarrow R$. Let's construct the truth table for our compound statement $((P \rightarrow Q) \wedge (Q \rightarrow R)) \rightarrow (P \rightarrow R)$.

Truth table for $((P \rightarrow Q) \wedge (Q \rightarrow R)) \rightarrow (P \rightarrow R)$

P	Q	R	$P \rightarrow Q$	$Q \rightarrow R$	$((P \rightarrow Q) \wedge (Q \rightarrow R))$	$P \rightarrow R$	$((P \rightarrow Q) \wedge (Q \rightarrow R)) \rightarrow (P \rightarrow R)$
T	T	T	T	T	T	T	T
T	T	F	T	F	F	F	T
T	F	T	F	T	F	T	T
T	F	F	F	T	F	F	T
F	T	T	T	T	T	T	T
F	T	F	T	F	F	T	T
F	F	T	T	T	T	T	T
F	F	F	T	T	T	T	T

Looking at the truth values for $((P \rightarrow Q) \wedge (Q \rightarrow R)) \rightarrow (P \rightarrow R)$, you see nothing but Ts. That's a great report card for a compound statement, and as a result, it gets put on the truth table equivalent of the honor roll. It is a tautology, and hence a valid argument. So "implies" has a transitive property.

You can apply this transitive property several times, as you will see in later examples.

Proof by Contradiction: The Advantage of Being Indirect

You have already observed that a conditional statement is logically equivalent to its contrapositive. It is reasonable to expect that if one logical version of a statement is a tautology, so is another. This is an important point because there are times when writing a direct proof to a theorem is awkward and difficult. When that happens, you might have an easier time if you turn to the negative side.

For example, suppose that if the power fails, you will be late for work. What conclusion can you draw if you were not late for work? For one thing, you know that the power must not have failed. Why? Well, the statement "the power fails" is either true or false (there is no other option). If the statement "the power fails" was true, then by your premise you would have been late for work. Because you were not late for work, it must be false that the power failed. This type of argument is called an indirect argument. You managed to make your argument through the back door, so to speak. You assumed an opposite conclusion, and contradicted your premises.

In an indirect proof of a theorem $P \to Q$, you assume $P \to Q$ and $\sim Q$, and try to show $\sim P$. You can write this as $((P \to Q) \wedge \sim Q) \to \sim P$, and construct a truth table for this compound statement.

Truth table for $((P \to Q) \wedge \sim Q) \to \sim P$

P	Q	$P \to Q$	$\sim Q$	$(P \to Q) \wedge \sim Q$	$\sim P$	$((P \to Q) \wedge \sim Q) \to \sim P$
T	T	T	F	F	F	T
T	F	F	T	F	F	T
F	T	T	F	F	T	T
F	F	T	T	T	T	T

Because $((P \to Q) \wedge \sim Q) \to \sim P$ is a tautology, you have yet another valid argument: Assume $P \to Q$ and $\sim Q$ and show $\sim P$.

Indirect proofs are particularly useful when proving statements about parallel lines. This is because when two lines are parallel, they do *not* intersect. This is a rather negative definition (the criteria for our definition is met by the two lines failing to do something in particular), and by negating the negative you turn everything positive. This can make things easier for you.

The Given Information: Use It or Lose It!

Whenever you set out to prove a theorem, it's vitally important that you use the given information. If you can prove a statement without using all of the given information, you should be suspicious. Mathematicians are not overly generous. They are happy to answer questions and share their knowledge, but they don't usually give more information than is needed or asked for.

When constructing a proof, it's best to start with the given information. That will serve to remind you of your assumptions. Somewhere along the way you will use premises that you either take on faith (postulates) or have previously proven (theorems).

A Solid Foundation: Definitions, Postulates, and Theorems

When you construct a proof, go step-by-step from your given information to (hopefully) your conclusion, or what you want to prove, using only your definitions, postulates, and theorems. The more theorems you have proven, the more sophisticated (and shorter) your proofs will become. You will build on what you've already established, and your house of cards will begin to stretch to the sky. As long as you start with a firm foundation of definitions and postulates, your structure will weather any storm. Guaranteed. It will be obvious if substandard parts are used, and any shoddy workmanship will be apparent immediately.

Write your proofs using the two-column technique. This technique has been used for centuries (or at least since I first learned geometry). It will help you organize your proofs and make them easy to read and understand. In the first column you write what you are claiming, step-by-step. In the second column you write why you can make that claim. Sometimes the justification for your claim is that you were given the information. In that case, all you need to include in your justification is "given." In other cases, your justification will be a definition, a postulate, or a theorem. If your postulate or theorem was given a name (like the Angle Addition Postulate), all you have to write down is the name. If your theorem or postulate wasn't special enough to warrant a name, you can either write down its number (they will all have numbers), or you can summarize the theorem. As for definitions, it's enough just to give the term whose definition you are using.

Let's prove a theorem using this two-column approach.

> **Theorem 8.1**: If $\overleftrightarrow{AC}$ intersects $\overleftrightarrow{BD}$ at O (as in Figure 8.1), then $\angle AOB \cong \angle DOC$.
>
> Proof: Here is your first look at a two-column proof in action.

	Statements	Reasons
1.	$\overleftrightarrow{AC}$ intersects $\overleftrightarrow{BD}$ at O	Given
2.	$\angle AOC$ and $\angle DOB$ are straight angles, and $m\angle AOC = 180°$, $m\angle DOB = 180°$	Definition of straight angle
3.	$m\angle AOC = m\angle DOB$	Substitution (step 2)
4.	$m\angle DOA + m\angle AOB = m\angle DOB$ and $m\angle DOA + m\angle DOC = m\angle AOC$	Angle Addition Postulate
5.	$m\angle DOA + m\angle AOB = m\angle DOA + m\angle DOC$	Substitution (steps 3 and 4)
6.	$m\angle AOB = m\angle DOC$	Subtraction property of equality
7.	$\angle AOB \cong \angle DOC$	Definition of congruence

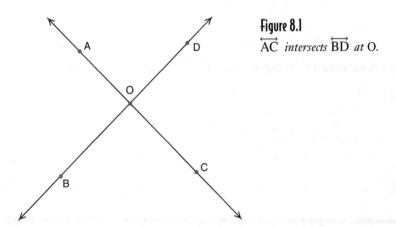

Figure 8.1

$\overleftrightarrow{AC}$ *intersects* $\overleftrightarrow{BD}$ *at* O.

Notice that in my proof I began with the given information. I used only one line to discuss the size of $\angle AOC$ and $\angle DOB$, because my reason for both statements was the same: the definition of a straight angle. Usually you only want to write one reason on each line. If the same reason is used to draw conclusions about two separate angles,

Eureka!

Theorem 8.1 is often stated as "When two lines intersect, the vertical angles formed are congruent."

line segments, and so on, then it's okay to put both conclusions on the same line, and list the reason only once. There's a certain amount of flexibility involved in writing a proof. As you become more familiar with formulating arguments you'll find yourself taking shortcuts with proof writing. Some shortcuts are fine, others are forbidden. I'll point out the shortcuts along the way, and also show you some common pitfalls to avoid.

What Should You Bring to a Formal Proof?

A formal proof of a statement is a sequence of steps that links the hypotheses of the statement to the conclusion of the statement using only deductive reasoning. The hypotheses and conclusion are usually stated in general terms. For example, you have already learned that when two lines intersect, the vertical angles formed are congruent. This statement is vague in that it does not name the specific lines that intersect or give the point of intersection. The statement does not even name the vertical angles formed. Because everything is so general, it's hard to get a handle on where to start. That's why in geometry a picture is worth a thousand words. When you draw a picture illustrating what's going on, it's easier to see what you are assuming and what you are trying to prove. You can translate your general statement into the picture in Figure 8.1. Here you have two intersecting lines, $\overleftrightarrow{AC}$ and $\overleftrightarrow{BD}$. They intersect at O. They form a couple of vertical angles, which you've named ∠AOB and ∠DOC. Your mission is to show that these two angles are congruent.

By drawing a picture, you've translated the general statement of the theorem into a specific example that you can pick apart and analyze. Try and include all of your assumptions in your picture. It's helpful to have your picture as complete as possible before beginning your proof, but that's not always possible. Sometimes as you work through a problem you realize that you need to label more things. You can always add labels to your pictures. Your diagram doesn't need to be set in stone until the proof is done.

After you've drawn your picture and labeled the important points, segments, lines, and so on, you have to interpret the hypotheses in terms of your picture. For the vertical angle example, you are given two lines that intersect. In your picture, the two intersecting lines are $\overleftrightarrow{AC}$ and $\overleftrightarrow{BD}$. In mathematical terms, you write that $\overleftrightarrow{AC}$ intersects $\overleftrightarrow{BD}$ at O.

The next step is to interpret what you would like to prove in terms of your picture. The general conclusion is that vertical angles are congruent. In your picture, the vertical angles are $\angle AOB$ and $\angle DOC$. You will prove that $\angle AOB \cong \angle DOC$.

The last step (and possibly the most difficult of them all) is to start with the given information and prove what you want to prove. This is done by thinking about your definitions, postulates, and any previous theorems that you have established. You want to use deductive reasoning to connect the given information to the conclusion.

Solid Facts _____

A **formal proof** has a definite style and format consisting of five essential elements.

♦ *Statement*. This states the theorem to be proved.

♦ *Drawing*. This represents the hypothesis of the theorem. Sometimes you have to translate the statement of the theorem into the specifics of your drawing.

♦ *Given*. This interprets the hypothesis of the theorem in terms of your drawing.

♦ *Prove*. This interprets the conclusion of the theorem in terms of your drawing.

♦ *Proof*. This orders a list of statements and justifications beginning with the given information and ending with what you wanted to prove. There must be a logical flow in this proof.

Let's practice writing a formal proof by proving Theorem 8.2.

> **Theorem 8.2**: If two lines intersect to form consecutive congruent angles, then these lines are perpendicular.

1. State the theorem. The theorem you will set out to prove is that if two lines intersect, and the consecutive angles are congruent, then these lines are perpendicular.

2. Draw a picture. Figure 8.2 will help us visualize what is going on. We have two lines, $\overleftrightarrow{AB}$ and $\overleftrightarrow{CD}$, that intersect at O. We have four angles to draw: $\angle 1$, $\angle 2$, $\angle 3$, and $\angle 4$. Notice that $\angle 1$ and $\angle 2$ are consecutive angles.

Figure 8.2

$\overleftrightarrow{AB}$ *and* $\overleftrightarrow{CD}$ *intersect at* O.

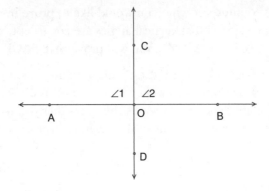

3. Given: $\overleftrightarrow{AB}$ and $\overleftrightarrow{CD}$ intersect at O, and $\angle 1 \cong \angle 2$.

4. Prove: $\overleftrightarrow{AB} \perp \overleftrightarrow{CD}$.

5. Write the proof.

Before you write out your columns, think about what it means for two lines to be perpendicular. Well, $\overleftrightarrow{AB} \perp \overleftrightarrow{CD}$ means that the two lines meet to form a right angle. So you need to show that when $\overleftrightarrow{AB}$ and $\overleftrightarrow{CD}$ intersect one right angle is formed. It doesn't matter which one is right; if one of them is right, they are all right. You are given that $\angle 1 \cong \angle 2$, and from your drawing you see that $\angle 1$ and $\angle 2$ are supplementary. So a little algebra will show that $m\angle 1 = 90°$, and hence is a right angle. So by the definition of perpendicular lines, $\overleftrightarrow{AB} \perp \overleftrightarrow{CD}$. When you have your drawing, you will want to look at it to come up with a game plan. Then you can put some columns down and write your proof.

	Statements	**Reasons**
1.	$\overleftrightarrow{AB}$ intersects $\overleftrightarrow{CD}$ at O, and $\angle 1 \cong \angle 2$	Given
2.	$m\angle 1 = m\angle 2$	Definition of $\cong$
3.	$\angle AOB$ is a straight angle, and $m\angle AOB = 180°$	Definition of straight angle
4.	$m\angle 1 + m\angle 2 = m\angle AOB$	Angle Addition Postulate
5.	$m\angle 1 + m\angle 2 = 180°$	Substitution (steps 3 and 4)
6.	$m\angle 1 + m\angle 1 = 180°$	Substitution (steps 2 and 5)
7.	$2m\angle 1 = 180°$	Algebra
8.	$m\angle 1 = 90°$	Algebra
9.	$\angle 1$ is right	Definition of right angle
10.	$\overleftrightarrow{AB} \perp \overleftrightarrow{CD}$	Definition of perpendicular lines

The first four steps involved in writing a formal proof deal mainly with reading the statement, drawing a picture, and interpreting what is given and what you are trying to prove in terms of your diagram. It's the last step that will require some thought. Before you draw your columns, think through what you are given. Try to imagine which definitions, postulates, or theorems would be useful in connecting the given information with the conclusion. When you have a game plan, everything should fall into place.

Remember to always start your proof with the given information, and end your proof with what you set out to show. As long as you do that, use one reason at a time, and only use definitions, postulates, and other theorems for your reasons, your proofs will flow like a mountain stream.

Put Me in, Coach!

Here's your chance to shine. Remember that I am with you in spirit and have provided the answers to these questions in Appendix A.

1. Is the following argument valid?

    a) If an angle is a right angle, then it measures 90°.

    b) ∠A is a right angle.

    Therefore m∠A = 90°.

2. Write a formal proof of the following theorem:

    Theorem 8.3: If two angles are complementary to the same angle, then these angles are congruent.

The Least You Need to Know

- ◆ A direct proof is structured using the compound statement $((P \rightarrow Q) \wedge P) \rightarrow Q$. We assume $P \rightarrow Q$ and P and try to conclude Q.

- ◆ An indirect proof is structured using the compound statement $((P \rightarrow Q) \wedge \sim Q) \rightarrow \sim P$.

- ◆ There are five essential components to a formal proof: statement, drawing, given, prove, and the proof.

Proving Segment and Angle Relationships

In This Chapter

- ◆ You can't have too many midpoints
- ◆ The one and only angle bisector
- ◆ Back up your claims about complements
- ◆ Establish some properties of supplements

You're ready to start making claims about segments and angles. Remember that you can be asked at any time to put your money where your mouth is and prove that what you say is true. Because mathematicians never exaggerate about the one that got away, there will be plenty of evidence to support your statements and persuade any skeptic to buy your claims.

Your claims will be your theorems, and you will back every claim with a two-column proof. In this chapter, I will walk you through each of the five steps needed to write a formal proof. I will repeat myself over and over and over again, not because I like to type the same words repeatedly, but so that writing proofs becomes second nature. You will reach a point when you will become psychic … you'll know what's next before you turn the page. When that happens (or at least by the time you finish reading this chapter), you will be able to crank out a formal proof like you've been doing it all of your life.

I'll start by establishing some facts about midpoints, and then move into the realm of angles. You'll get a chance to review some postulates and definitions, and move closer to becoming good friends with algebra.

Exploring Midpoints

Recall that the midpoint of $\overline{AB}$ is a point M on $\overline{AB}$ that divides $\overline{AB}$ into two congruent pieces. You can use this definition to prove that each piece has length $\frac{1}{2}$ AB. That's certainly reasonable. If you divide a segment into two pieces of equal length, it makes sense that half of the original length will go to the first piece, the other half to the second piece. This is such a reasonable statement, it's just got to be a theorem. Consider this your first invitation to a formal proof. I'll go through each of the five steps in the process.

> **Example 1**: Prove that the midpoint of a segment divides the segment into two pieces, each of which has length equal to one-half the length of the original segment.

> **Solution**: Follow the steps outlined in how to write a formal proof.

> 1. Give a statement of the theorem:

> **Theorem 9.1**: The midpoint of a segment divides the segment into two pieces, each of which has length equal to one-half the length of the original segment.

> 2. Draw a picture. Theorem 9.1 talks only about a line segment and its midpoint. So Figure 9.1 only shows $\overline{AB}$ with midpoint M.

Figure 9.1

M *is the midpoint of* $\overline{AB}$.

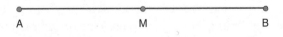

> 3. State what is given in terms of our drawing. Given: a line segment $\overline{AB}$ and a midpoint M.

> 4. State what you want to prove in terms of your drawing. Prove: AM $= \frac{1}{2}$ AB.

> 5. Write the proof. You need a game plan. In proving this theorem, you will want to make use of any definitions, postulates, and theorems that you have at your disposal. The definition you will want to use is that of the midpoint.

The postulate that will come in handy is the Segment Addition Postulate, which states that if X is a point on $\overline{AB}$, then $AX + XB = AB$. This theorem doesn't seem to have any special needs, so you will prove this theorem directly. Start with your given information, and don't stop until $AM = \frac{1}{2} AB$.

	Statements	Reasons
1.	M is the midpoint of $\overline{AB}$	Given
2.	$\overline{AM} \cong \overline{MB}$	Definition of midpoint
3.	$AM = MB$	Definition of $\cong$
4.	$AM + MB = AB$	Segment Addition Postulate
5.	$2AM = AB$	Substitution (steps 3 and 4)
6.	$AM = \frac{1}{2} AB$	Algebra

There is a little flexibility in the reasons given, especially when you are dealing with algebra. For example, the reason for step 6 was "algebra," but it could also have been "the division property of equality." When it comes to the geometrical parts of a proof, however, there is not much flexibility. The reason for step 2 could only have been the definition of a midpoint, and step 3 is valid only because of the Segment Addition Postulate. There are no other options in these cases.

How Many Midpoints Are There?

Mathematicians talk about the midpoint of a segment as if it is the only one. It makes sense that there is only one "middle" of a segment, but suppose someone has just challenged you to prove it. Not only will you get practice writing proofs, but you'll get to dust off your algebra skills and put them to work as well.

There are a couple of ways that you can prove that a line segment can only have one midpoint. You can try the direct approach: Start with a line segment, find the midpoint and show that no other point on the segment has what it takes to be a midpoint. This approach is difficult. You would have to examine each point on the segment for its midpoint potential. Unfortunately, there are infinitely many points on a line segment. Even if you dedicated the rest of your life to completing this process, it would not be enough.

In this situation a direct approach is much more difficult than an indirect approach. To use an indirect approach, turn to the negative of your conclusion. The conclusion is that the midpoint is unique, or that there is only one midpoint. The negative of that statement is that the midpoint is not unique. There is not only one. That means that there must be at least two midpoints.

This method of an indirect proof is often referred to as a "proof by contradiction." You start by assuming that the conclusion is false (in this case, that the midpoint is not unique), and try to contradict one of your definitions, postulates, theorems or assumptions. Let's see how it all plays out.

Example 2: Prove that the midpoint of a segment is unique.

Solution: Go through the five steps involved in writing a formal proof.

1. Give a statement of the theorem:

Theorem 9.2: The midpoint of a segment is unique.

2. Create a drawing to visualize what's going on. You'll need a line segment; call it $\overline{AB}$. You'll also need two midpoints, which you can call M and N. I've drawn everything you'll need in Figure 9.2.

Figure 9.2

$\overline{AB}$ *has two distinct mid-points* M *and* N

3. State what is given in terms of the drawing. You are given a line segment $\overline{AB}$ and two distinct midpoints M and N.

4. State what you want to prove in terms of the drawing. According to your drawing, the line segment $\overline{AB}$ has been broken up into three segments: $\overline{AM}$, $\overline{MN}$, and $\overline{NB}$. You don't have many definitions or theorems about segments that you can contradict, but there is the Ruler Postulate: The measure of any line segment is a unique *positive* number. If you can somehow show that MN = 0, you will contradict this postulate. So that's what you'll do. Prove: MN = 0.

5. Write the proof. Your game plan is to somehow argue that MN = 0. You can only use the fact that *both* M and N are midpoints, and use Theorem 9.1. It's going to take some algebra, but you can do it. Put down your columns and walk through the argument step-by-step. Remember to start with the given information, and stop when you have shown that MN = 0.

	Statements	Reasons
1.	M and N are both midpoints of $\overline{AB}$	Given
2.	$AM = \frac{1}{2}AB$ and $AN = \frac{1}{2}AB$	Theorem 9.1
3.	$AM + MN + NB = AB$	Segment Addition Postulate

	Statements	Reasons
4.	$\frac{1}{2}AB + MN + \frac{1}{2}AB = AB$	Substitution (steps 2 and 3)
5.	$AB + MN = AB$	Algebra
6.	$MN = 0$	Subtraction property of equality

Proving Angles Are Congruent

Two angles are congruent if they have the same measure. You already know that when two lines intersect the vertical angles formed are congruent. You have also seen that if $\angle A$ and $\angle B$ are each complementary to $\angle C$, then $\angle A \cong \angle B$. There are other angle relationships to explore. When you expose these angle relationships, you will establish their truth using a formal proof.

For example, you were introduced to the idea of an angle bisector in Chapter 4. Well, it turns out that the bisector of an angle divides the angle into two angles, each of which has measure equal to one-half the measure of the original angle.

This statement looks a lot like Theorem 9.1 applied to angles rather than segments. You can use a game plan similar to the one you used to prove Theorem 9.1 to prove this theorem.

> **Example 3**: Prove that the bisector of an angle divides the angle into two angles, each of which has measure equal to one-half the measure of the original angle.

> **Solution**: Go step-by-step through the formal proof.

1. Give a statement of the theorem.

Theorem 9.3: The bisector of an angle divides the angle into two angles, each of which has measure equal to one-half the measure of the original angle.

2. Draw a picture. You need an angle and its bisector. Figure 9.3 shows $\angle ABC$ bisected by $\overrightarrow{BD}$.

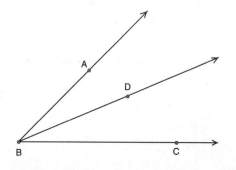

Figure 9.3

$\angle ABC$ *is bisected by* $\overrightarrow{BD}$.

3. State what is given in terms of your drawing. You are given ∠ABC which is bisected by $\overrightarrow{BD}$.

4. State what you want to prove in terms of your drawing. You want to prove that m∠ABD = ½ m∠ABC.

5. Write the proof. You must think about which definitions, postulates, and theorems you can make use of. The first one that comes to mind is the definition of an angle bisector. The postulate that will come in handy is the Angle Addition Postulate, which states that if a point D lies in the interior of ∠ABC, then m∠ABD + m∠DBC = m∠ABC . That's most of what you'll need to cook up this proof. You'll need a pinch of algebra to complete the dish.

	Statements	Reasons
1.	$\overrightarrow{BD}$ is the angle bisector of ∠ABC	Given
2.	∠ABD ≅ ∠DBC	Definition of angle bisector
3.	m∠ABD = m∠DBC	Definition of
4.	m∠ABD + m∠DBC = m∠ABC	Angle Addition Postulate
5.	2m∠ABD = m∠ABC	Substitution (steps 3 and 4)
6.	m∠ABD = ½ m∠ABC	Algebra

Using and Proving Angle Complements

There are lots of relationships between angles that can be proven formally. For example, there are many situations where two seemingly unrelated angles can be shown to be complements of each other. Recall that two angles are complementary if the sum of their measures is 90°. For example, in Figure 9.4 there are several angles hanging around. There's ∠ABC, ∠CBD, and ∠DBE. Suppose that ∠CBD is a right angle. Then ∠ABC and ∠DBE are complementary. I'll write an informal proof of this.

Figure 9.4

∠ABE is straight and ∠CBD is a right angle.

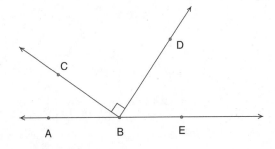

Example 4: Prove that if, as shown in Figure 9.4, ∠ABE is straight and ∠CBD is a right angle, then ∠ABC and ∠DBE are complementary.

Solution: All we need for an informal proof is a picture, the columns, and a game plan.

Here's the game plan: You are going to want to break apart your angles, so you know that the Angle Addition Postulate will be useful. You will need the definition of a straight angle, a right angle, and the definition of complementary angles in order to translate your angle characteristics into numbers. Then you'll be ready to use some algebra to bring it home.

	Statements	Reasons
1.	∠ABE is straight and ∠CBD is a right angle	Given
2.	m∠ABE = 180°	Definition of straight angle
3.	m∠CBD = 90°	Definition of right angle
4.	m∠ABC + m∠CBD + m∠DBE = m∠ABE	Angle Addition Postulate
5.	m∠ABC + 90° + m∠DBE = 180°	Substitution (steps 2, 3, 4)
6.	m∠ABC + m∠DBE = 90°	Algebra
7.	∠ABC and ∠DBE are complementary.	Definition of complementary angles

You started with your given information, used the definitions of the terms involved in the statement of the theorem, and finished up with what you wanted to prove. Every step has a reason that is either a definition, a postulate, an already-established theorem, or algebra. And the proof only required seven steps to write!

When I first introduced you to the concept of an angle, I threw out several angle relationships and gave brief explanations about why my claims were reasonable. Being the agreeable type of reader that you are, you didn't question me (or if you did, I didn't hear you). I'll take a minute and address one of the statements I made that might have raised an eyebrow or two. The statement in question is "the complement of an acute angle is an acute angle." There's no time like the present to write a formal proof of this potentially bold claim.

Example 5: Write a formal proof that the complement of an acute angle is an acute angle.

Solution: Let's try a new approach. Let's work through the five steps in writing a proof. (Note the sarcasm.)

1. State the theorem.

Theorem 9.4: The complement of an acute angle is an acute angle.

2. Draw a picture. You need two angles, one of which is acute, whose measures add up to 90°. In other words, the two angles must combine to form a right angle. I have drawn these two angles in Figure 9.5.

Figure 9.5

An acute angle ∠ABC and its complement, ∠CBD.

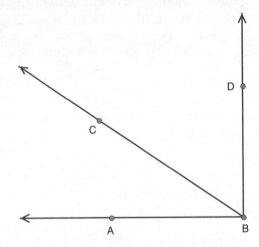

3. Interpret what is given in terms of your picture. You are given an acute angle ∠ABC, and its complement ∠CBD.

4. Interpret what you want to prove in terms of your drawing. You want to prove that ∠CBD is acute.

5. Write the proof. Again, you need a game plan. You are given an acute angle, so the definition of an acute angle will be useful. Because you are combining angles, you might want to use the Angle Addition Postulate. Because you are trying to show that ∠CBD is acute, you need to show that m∠CBD < 90°. So you'll need to refresh what it means for m∠CBD to be less than 90°. Recall that x is less than y if there exists a positive number z for which x + y = z. Because the measure of an angle is a positive number between 0 and 180 (recall the Protractor Postulate), you should be all set.

	Statements	Reasons
1.	∠ABC is acute, and ∠ABC and ∠CBD are complementary.	Given
2.	m∠ABC + m∠CBD = 90°	Definition of complementary angles

	Statements	Reasons
3.	m∠ABC > 0	Protractor Postulate
4.	m∠CBD < 90°	Definition of inequality
5.	∠CBD is acute	Definition of acute angle

Is it my imagination, or are these proofs getting shorter and easier to write?

Using and Proving Angle Supplements

When acute angles need a supplement, they turn to an obtuse angle. Even though I am an authority on the subject, and I make that statement with a certain level of confidence and conviction, skeptics should demand that I put my money where my mouth is and prove it. And to set a good example, I will.

Example 6: Prove that the supplement of an acute angle is an obtuse angle.

Solution: You know what to do, so just do it.

1. State the theorem.

Theorem 9.5: The supplement of an acute angle is an obtuse angle.

2. Draw a picture. Figure 9.6 shows an acute angle ∠ABC and its supplement ∠CBD. Together ∠ABC and ∠CBD form the straight angle ∠ABD.

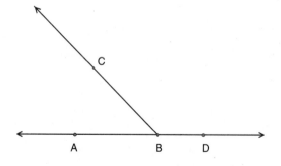

Figure 9.6

∠ABC *and* ∠CBD *are supplementary angles.*

3. Interpret what you are given in terms of your drawing. You are given ∠ABC and its supplement ∠CBD, with ∠ABC acute.

4. Interpret what you are trying to prove in terms of your drawing. Prove that ∠CBD is obtuse.

5. Prove the theorem. What's the game plan? This proof involves acute, obtuse, and supplementary angles, so you'll probably use their definitions somewhere. Because you'll be dealing with inequalities (acute angles have measure less than 90° and obtuse angles have measure greater than 90°), you might need your definitions of < or >, and you might need our Protractor Postulate. And there's always algebra.

	Statements	Reasons
1.	∠ABC is acute, and ∠ABC and ∠CBD are supplementary.	Given
2.	m∠ABC + m∠CBD = 180°	Definition of supplementary angles
3.	m∠ABC < 90°	Definition of acute angle
4.	m∠ABC + 90° < 180°	Addition property of inequality
5.	m∠ABC + 90° < m∠ABC + m∠CBD	Substitution (steps 2 and 4)
6.	90° < m∠CBD	Subtraction property of inequality
7.	∠CBD is obtuse	Definition of obtuse angle

Now that you're starting to crank out those formal proofs, it's time to open things up and see how you perform on the open road. Take a look at Figure 9.7. ∠ABC and ∠CBD are adjacent supplementary angles. If you construct the bisectors of each of these two angles, then together the bisectors will form a new angle. But not just any angle. This new angle will be right. And you'll prove it.

Figure 9.7

∠ABC *and* ∠CBD *are adjacent supplementary angles;* $\overrightarrow{BE}$ *bisects* ∠ABC, *and* $\overrightarrow{BF}$ *bisects* ∠CBD.

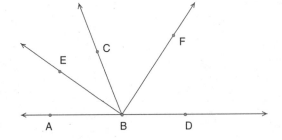

Example 7: Prove that the bisectors of two adjacent supplementary angles form a right angle.

Solution: Follow these steps.

1. State the theorem.

Theorem 9.6: The bisectors of two adjacent supplementary angles form a right angle.

2. Draw a picture (see Figure 9.7).

3. Interpret the given information in terms of the picture. $\angle ABC$ and $\angle CBD$ are adjacent supplementary angles; $\overrightarrow{BE}$ bisects $\angle ABC$, and $\overrightarrow{BF}$ bisects $\angle CBD$.

4. Interpret what to prove in terms of the picture. Prove that $\angle EBF$ is a right angle.

5. Prove the theorem. Your game plan: You'll need some definitions in this proof: supplementary angles, right angles, and angle bisectors. Because you will be breaking up angles, the Angle Addition Postulate might be useful. Let's see how it all unfolds.

	Statements	Reasons
1.	$\angle ABC$ and $\angle CBD$ are adjacent supplementary angles; $\overrightarrow{BE}$ bisects $\angle ABC$, and $\overrightarrow{BF}$ bisects $\angle CBD$	Given
2.	$\angle ABE \cong \angle EBC$, $\angle CBF \cong \angle FBD$	Definition of angle bisector
3.	$m\angle ABE = m\angle EBC$, $m\angle CBF = m\angle FBD$	Definition of $\cong$
4.	$m\angle ABC + m\angle CBD = 180°$	Definition of supplementary angles
5.	$m\angle ABE + m\angle EBC = m\angle ABC$, $m\angle CBF + m\angle FBD = m\angle CBD$, and $m\angle EBC + m\angle CBF = m\angle EBF$	Angle Addition Postulate
6.	$m\angle ABE + m\angle EBC + m\angle CBF + m\angle FBD = 180°$	Substitution (steps 4 and 5)
7.	$2m\angle EBC + 2m\angle CBF = 180°$	Substitution (steps 3 and 6)
8.	$m\angle EBC + m\angle CBF = 90°$	Algebra
9.	$m\angle EBF = 90°$	Substitution (steps 5 and 8)
10.	$\angle EBF$ is right	Definition of right angle

Keep in mind that there is more than one way to construct a proof. If you put three mathematicians in a room and have them prove the same theorem, you will probably get three different proofs. They would all be valid (assuming they did it right), though they might have taken different steps along the way. Variety is the spice of life. Just be sure to avoid using cheap reasons in your proofs. Trust me: It will show.

Put Me in, Coach!

Here's your chance to shine. Remember that I am with you in spirit and have provided the answers to these questions in Appendix A.

1. If E is between D and F, write a formal proof that DE = DF – EF.

2. Given ∠ABC and $\overrightarrow{BD}$ as in Figure 9.8, write a formal proof that
   m∠ABD = m∠ABC – m∠DBC .

Figure 9.8
$\overrightarrow{BD}$ *divides* ∠ABC
into two angles

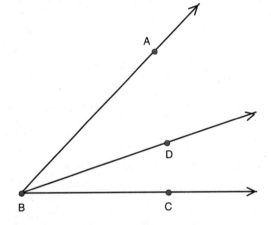

3. Prove that the angle bisector of an angle is unique.

4. Prove that the supplement of a right angle is a right angle.

The Least You Need to Know

♦ The midpoint of a line segment is unique.

♦ The angle bisector of an angle is unique.

♦ The complement of an acute angle is an acute angle.

♦ The supplement of an acute angle is an obtuse angle.

♦ The bisectors of two adjacent supplementary angles form a right angle.

♦ Writing a two-column proof gets easier the more you practice writing them.

10

Proving Relationships Between Lines

In This Chapter

- ◆ How corresponding angles correspond
- ◆ Another shot at writing an indirect proof
- ◆ How parallel lines determine angle relationships
- ◆ Use angle relationships to show lines are parallel

Equivalence relations are pretty important in mathematics. The usefulness of = cannot be overstated, and we've used ≅ in many of our formal proofs. You might begin to think that the only relations worth talking about are equivalence relations. That would be too simple.

You know that the relations ⊥ and || are not equivalence relations. Nevertheless, these two relations are proof that you don't have to be an equivalence relation to be worthy of further study. Without a perpendicularity you could never build a kite. Without a parallelism you would never be able to set a glass down on a table without worrying about it sliding off and ruining the carpet.

A lot can happen when two parallel lines are cut by a transversal. Some very important angle relationships are just waiting to be discovered. In fact, there are some angle relationships strong enough to force two lines to be parallel. I'll start with the perpendicular, and move into the realm of the parallel.

Proofs Involving Perpendicular Lines

I'll begin with a review of what you've learned about lines. Whenever you have two lines, only one of three things can happen: Either they are the same line, they are parallel lines, or the two lines intersect at a point. If the two lines intersect at a point, the vertical angles formed are congruent. The intersecting lines either form a pair of acute angles and a pair of obtuse angles, or the intersecting lines form four right angles. When the lines meet to form four right angles, the lines are perpendicular.

The main fact to establish about perpendicular lines has to do with uniqueness. Remember that that the midpoint of a line segment and the angle bisector of an angle are unique. In Chapter 5 you learned that if you are given a point and a line, there is a unique line passing through that point that is perpendicular to the line. You now have the skills to establish the uniqueness property of perpendicular lines.

> **Theorem 10.1**: Given a point A on a line l, there exists a unique line m perpendicular to l which passes through A.

> **Example 1**: Write a formal proof for Theorem 10.1.

> **Solution**: Start with a game plan for how to approach the problem. Figure 10.1 shows a line l and a point A on l. You want to show that there is a unique line m perpendicular to l which passes through A. The way you proved uniqueness in earlier examples was to assume that there were two, and obtain a contradiction. That's the same approach to take here.

Figure 10.1

A line l *and a point* A *on* l.

The drawing you will use for your proof needs two distinct lines, m and n, which both pass through A and are perpendicular to l. Figure 10.2 illustrates that situation. The contradiction you'll obtain involves the Protractor Postulate. Recall that when two lines are perpendicular, they meet to form right angles. Lines m and l form $\angle 3$. Lines n and l form $\angle 2$. Because m and n are distinct lines that meet at A, when they intersect they will form $\angle 1$. Together $\angle 1$, $\angle 2$ and $\angle 3$ form the straight angle $\angle BAC$, so the sum of their measures must be 180°. But if $m\angle 2 = 90°$ and $m\angle 3 = 90°$, you have accounted for all of the 180°. There are no more degrees left over to form $\angle 1$. That's where the problem lies: $m\angle 1 = 0°$, which contradicts the Protractor Postulate. Now that you have a game plan, you can write the formal proof. At this point you should be comfortable with the format of a formal proof, so I'll just go through the steps.

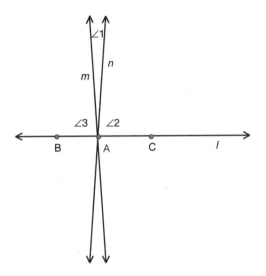

Figure 10.2

Two distinct lines, m and n, which both pass through A and are perpendicular to l.

Theorem 10.1: Given a point A on a line l, there exists a unique line m perpendicular to l which passes through A.

The drawing is shown in Figure 10.2.

Given a line l and a point A on l, suppose there are two lines, m and n, which both pass through A and are perpendicular to l.

Prove that $m\angle 1 = 0°$

Proof: As far as a game plan goes, I have already outlined most of the proof. You'll use the definition of a straight angle, the Angle Addition Postulate, and the Protractor Postulate.

	Statements	Reasons
1.	Points A, B, and C lie on a line l, and m and n are distinct lines which both pass through A and are perpendicular to l	Given
2.	$\angle$BAC is a straight angle, and $m\angle$BAC $= 180°$	Definition of straight angle
3.	$m\angle 1 + m\angle 2 + m\angle 3 = m\angle$BAC	Angle Addition Postulate
4.	$m\angle 1 + m\angle 2 + m\angle 3 = 180°$	Substitution (steps 2 and 3)

	Statements	Reasons
5.	$\angle 2$ is a right angle	Definition of perpendicular ($n \perp l$)
6.	$\angle 3$ is a right angle	Definition of perpendicular ($m \perp l$)
7.	$m\angle 2 = 90°$, $m\angle 3 = 90°$	Definition of right angle
8.	$m\angle 1 + 90° + 90° = 180°$	Substitution (steps 4 and 7)
9.	$m\angle 1 = 0°$	Algebra

You have established your contradiction, and thus the assumption that there were two distinct lines perpendicular to l passing through A was false. Uniqueness is established.

Let's Get Parallel

You were introduced to Euclid's Parallel Postulate in Chapter 5. That postulate has raised more questions in the last 2,000 years than any other postulate. Euclid spent a lot of time pondering parallel lines. He was very uncomfortable with his 5th postulate, because he firmly believed that he could prove it from other postulates and theorems. In this section, you'll be dealing with pairs of lines cut by a transversal. Remember all of the angle relationships that were involved with lines and transversals. There are alternate interior and exterior angles, same-side interior and exterior angles, corresponding angles, and vertical angles. Feel free to flip back to Chapter 5 and review these topics if you need some refreshment.

When dealing with parallel lines, you can freely invoke Euclid's 5th postulate. It's important to keep in mind that many of these results only hold for Euclidean geometry. You'll have to wait until Chapter 24 to enter the bizarre world of non-Euclidean geometry.

In order to explore the relationships between angles formed when parallel lines are cut by a transversal, you need a place to start. It's time to add another postulate to your collection.

> **Postulate 10.1**: If two parallel lines are cut by a transversal, then the corresponding angles are congruent.

To get a feel for this theorem, take a look at Figure 10.3. If lines l and m are parallel, and line t is a transversal, then $\angle 1$ and $\angle 2$ are corresponding angles. According to Postulate 10.1, $\angle 1 \cong \angle 2$. We can use this postulate to solve "find the measure of the angle" puzzles.

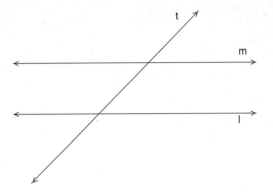

Figure 10.3

Lines l and m are parallel, line t is a transversal.

Example 2: In Figure 10.4, l‖m and m∠1 = 112°. Find m∠2, m∠4, m∠5, and m∠8.

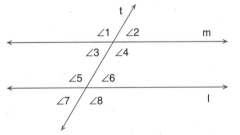

Figure 10.4

l‖m *and* m∠1 = 112°.

Solution: Use Postulate 10.1, and the theorem about vertical angles to solve this problem. Because ∠1 and ∠4 are vertical angles, you know that they are congruent. Right away you know that m∠4 = 112°. Because ∠4 and ∠8 are corresponding angles, by Postulate 10.1 you know that ∠4 ≅ ∠8. Because m∠4 = 112°, m∠8 = 112°. Because ∠8 and ∠5 are vertical angles, you know that ∠5 ≅ ∠8, so m∠5 = 112°. All that's left is to find m∠2. Because ∠1 and ∠2 are supplementary angles (together they form a straight angle), you know that m∠1 + m∠2 = 180°. Substituting in for m∠1, you see that m∠1 + 112° = 180°. So m∠2 = 68°. To summarize: m∠2 = 68°, m∠4 = 112°, m∠5 = 112°, and m∠8 = 112°.

Proofs About Alternate Angles

You can use Postulate 10.1 to prove lots of theorems. Most of these theorems will travel in pairs. Here are the first of many pairs of theorems.

Theorem 10.2: If two parallel lines are cut by a transversal, then the alternate interior angles are congruent.

Theorem 10.3: If two parallel lines are cut by a transversal, then the alternate exterior angles are congruent.

I'll write out a proof of Theorem 10.2 and give you the opportunity to prove Theorem 10.3 at the end of this chapter. To prove Theorem 10.2, you'll need a couple of parallel lines cut by a transversal, two alternate interior angles, and an angle that corresponds to one of those alternate interior angles. Figure 10.5 shows the important angles.

Figure 10.5

1∥ m *cut by a transversal* t.

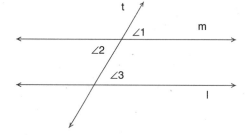

Given: 1∥ m cut by a transversal t.

Prove: ∠2 ≅ ∠3

Proof: The game plan is straightforward. ∠1 and ∠3 are corresponding angles, so they are congruent by Postulate 10.1. ∠1 and ∠2 are vertical angles, so you know that they are congruent. The transitive property of ≅ provides the rest.

	Statements	**Reasons**
1.	1∥ m cut by a transversal t	Given
2.	∠2 and ∠3 are alternate interior angles	Definition of alternate interior angles
3.	∠1 and ∠3 are corresponding angles.	Definition of corresponding angles
4.	∠1 and ∠2 are vertical angles	Definition of vertical angles
5.	∠1 ≅ ∠3	Postulate 10.1
6.	∠4 ≅ ∠8	Theorem 8.1
7.	∠2 ≅ ∠3	Transitive property of ≅

Parallel Lines and Supplementary Angles

Whenever two parallel lines are cut by a transversal, an interesting relationship exists between the two interior angles on the same side of the transversal. These two interior angles are supplementary angles. A similar claim can be made for the pair of exterior angles on the same side of the transversal. There are two theorems to state and prove. I'll give formal statements for both theorems, and write out the formal proof for the first. The second theorem will provide yet another opportunity for you to polish your formal proof writing skills.

> **Theorem 10.4**: If two parallel lines are cut by a transversal, then the interior angles on the same side of the transversal are supplementary angles.

> **Theorem 10.5**: If two parallel lines are cut by a transversal, then the exterior angles on the same side of the transversal are supplementary angles.

Let the fun begin. As promised, I will show you how to prove Theorem 10.4.

Figure 10.6 illustrates the ideas involved in proving this theorem. You have two parallel lines, l and m, cut by a transversal t. You will be focusing on interior angles on the same side of the transversal: $\angle 2$ and $\angle 3$. You'll need to relate to one of these angles using one of the following: corresponding angles, vertical angles, or alternate interior angles. There are many different approaches to this problem. Because Theorem 10.2 is fresh in your mind, I will work with $\angle 1$ and $\angle 3$, which together form a pair of alternate interior angles.

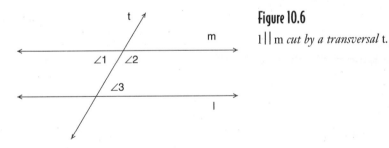

Figure 10.6

l || m *cut by a transversal* t.

Given: l || m cut by a transversal t.

Prove: $\angle 2$ and $\angle 3$ are supplementary angles.

Proof: You will need to use the definition of supplementary angles, and you'll use Theorem 10.2: When two parallel lines are cut by a transversal, the alternate interior angles are congruent. That should be enough to complete the proof.

	Statements	Reasons
1.	l‖m cut by a transversal t	Given
2.	∠2 and ∠3 are same-side interior angles	Definition of same-side interior angles
3.	∠1 and ∠3 are alternate interior angles	Definition of alternate interior angles
4.	∠1 and ∠2 are supplementary angles, and $m\angle 1 + m\angle 2 = 180°$	Definition of supplementary angles
5.	$\angle 1 \cong \angle 3$	Theorem 10.2
6.	$m\angle 1 = m\angle 3$	Definition of $\cong$
7.	$m\angle 3 + m\angle 2 = 180°$	Substitution (steps 4 and 6)
8.	∠2 and ∠3 are supplementary angles	Definition of supplementary angles

Using Parallelism to Prove Perpendicularity

Suppose you have the situation shown in Figure 10.7. Two lines, l and m, are parallel, and are cut by a transversal t. In addition, suppose that l⊥t. In this case, you can conclude that m⊥t. There are those who would doubt your conclusions, and it is for those people that I include a proof. As it is stated, the problem cannot have theorem status. Theorems are typically general statements, like "when two lines intersect, the vertical angles formed are congruent." In this case, your observation came from a specific situation, and it cannot become a theorem unless it is written in more general terms, like "when two parallel lines are cut by a transversal, if one of the lines is perpendicular to the transversal, then both of the lines are perpendicular to the transversal." That's the stuff that theorems are made of. Here's a formal proof of the theorem.

Figure 10.7

Lines l *and* m *are parallel lines cut by a transversal* t, *with* l⊥t .

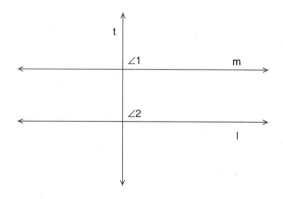

Theorem 10.6: When two parallel lines are cut by a transversal, if one of the lines is perpendicular to the transversal, then both of the lines are perpendicular to the transversal.

Figure 10.7 illustrates the situation nicely.

Given: Lines l and m are parallel and are cut by a transversal t, l⊥t .

Prove: m⊥t

Proof: Your game plan is to use Postulate 10.1, which says that when two parallel lines are cut by a transversal, corresponding angles are congruent. Because l and t meet to form a right angle, so will m and t, making them perpendicular.

	Statements	Reasons
1.	l‖m cut by a transversal t, l⊥t	Given
2.	∠1 is right	Definition of perpendicular
3.	m∠1 = 90°	Definition of right angle
4.	∠1 and ∠2 are corresponding angles	Definition of corresponding angles
5.	∠4 ≅ ∠8	Postulate 10.1
6.	m∠1 = m∠2	Definition of ≅
7.	m∠2 = 90°	Substitution (steps 2 and 5)
8.	∠2 is right	Definition of right angle
9.	m⊥t	Definition of ⊥

Proving Lines Are Parallel

When you were given Postulate 10.1, you were able to prove several angle relationships that developed when two parallel lines were cut by a transversal. There are times when particular angle relationships are given to you, and you need to determine whether or not the lines are parallel. You'll develop some theorems to help you do this easily. Your first theorem, Theorem 10.7, will be established using contradiction. The rest of the theorems will follow using a direct proof and Theorem 10.7.

Let's review the steps involved in constructing a proof by contradiction. Start by assuming that the conclusion is false, and then showing that the hypotheses must also be false. In the original statement of the proof, you start with congruent corresponding angles and conclude that the two lines are parallel. To prove this theorem using contradiction, assume that the two lines are not parallel, and show that the corresponding angles cannot be congruent.

Theorem 10.7: If two lines are cut by a transversal so that the corresponding angles are congruent, then these lines are parallel.

A drawing of this situation is shown in Figure 10.8. Two lines, l and m are cut by a transversal t, and ∠1 and ∠2 are corresponding angles.

Figure 10.8

l *and* m *are cut by a transversal* t, *and* ∠1 *and* ∠2 *are corresponding angles.*

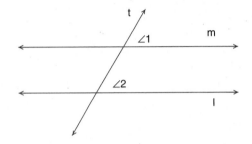

Given: l and m are cut by a transversal t, l ∦ m.

Prove: ∠1 and ∠2 are not congruent (∠1 ≇ ∠2).

Proof: Assume that l ∦ m. Because l and m are cut by a transversal t, m and t must intersect. You might call the point of intersection of m and t the point O. Because l is *not* parallel to m, we can find a line, say r, that passes through O and *is* parallel to l. I've drawn this new line in Figure 10.9. In this new drawing, ∠3 and ∠2 are corresponding angles, so by Postulate 10.1, they are congruent. But wait a minute! If ∠2 ≅ ∠3, and m∠3 + m∠4 = m∠1 by the Angle Addition Postulate, m∠2 + m∠4 = m∠1 . Because m∠4 > 0 (by the Protractor Postulate), this means that m∠2 < m∠1, and ∠1 ≇ ∠2 . Let's put this all down in two columns.

Figure 10.9

l *and* m *are cut by a transversal* t, l ∦ m, r ∥ l, *and* r, m, *and* l *intersect at* O.

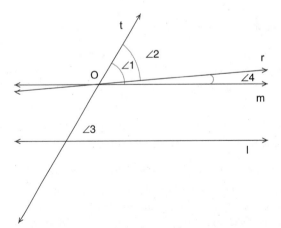

	Statements	Reasons
1.	l and m are two lines cut by a transversal t, with l ∦ m	Given
2.	Let r be a line passing through O which is parallel to l	Euclid's 5th postulate
3.	∠3 and ∠2 are corresponding angles	Definition of corresponding angles
4.	∠2 ≅ ∠3	Postulate 10.1
5.	m∠2 = m∠3	Definition of ≅
6.	m∠3 + m∠4 = m∠1	Angle Addition Postulate
7.	m∠2 + m∠4 = m∠1	Substitution (steps 5 and 6)
8.	m∠4 > 0	Protractor Postulate
9.	m∠2 < m∠1	Definition of inequality
10.	∠4 ≇ ∠8	Definition of ≅

That completes your proof by contradiction. The rest of the theorems that you prove in this section will make use of Theorem 10.7. The rest of the theorems in this chapter are converses of theorems proved earlier.

Let's take a look at some other angle relationships that can be used to prove that two lines are parallel. These two theorems are similar, and to be fair I will prove the first one and leave you to prove the second.

> **Theorem 10.8**: If two lines are cut by a transversal so that the alternate interior angles are congruent, then these lines are parallel.

> **Theorem 10.9:** If two lines are cut by a transversal so that alternate exterior angles are congruent, then these lines are parallel.

Figure 10.10 shows two lines cut by a transversal t, with alternate interior angles labeled ∠1 and ∠2.

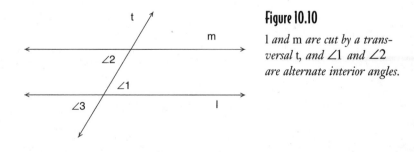

Figure 10.10

l *and* m *are cut by a transversal* t, *and* ∠1 *and* ∠2 *are alternate interior angles.*

Given: l and m are cut by a transversal t, with $\angle 4 \cong \angle 8$.

Prove: l || m.

Proof: The game plan is simple. In order to use Theorem 10.7, you need to show that corresponding angles are congruent. You can use the fact that $\angle 1$ and $\angle 2$ are vertical angles, so they are congruent. Because $\angle 2$ and $\angle 3$ are corresponding angles, if you can show that they are congruent, then you will be able to conclude that your lines are parallel. The transitive property of congruence will put the nail in the coffin, so to speak.

	Statements	Reasons
1.	l and m are two lines cut by a transversal t, with $\angle 4 \cong \angle 8$	Given
2.	$\angle 1$ and $\angle 3$ are vertical angles	Definition of vertical angles
3.	$\angle 1 \cong \angle 3$	Theorem 8.1
4.	$\angle 2$ and $\angle 3$ are corresponding angles	Definition of corresponding angles
5.	$\angle 2 \cong \angle 3$	Transitive property of $\cong$
6.	l || m	Theorem 10.7

Theorem 10.4 established the fact that if two parallel lines are cut by a transversal, then the interior angles on the same side of the transversal are supplementary angles. Theorem 10.5 claimed that if two parallel lines are cut by a transversal, then the exterior angles on the same side of the transversal are supplementary angles. It's now time to prove the converse of these statements. Let's split the work: I'll prove Theorem 10.10 and you'll take care of Theorem 10.11.

Theorem 10.10: If two lines are cut by a transversal so that the interior angles on the same side of the transversal are supplementary, then these lines are parallel.

Theorem 10.11: If two lines are cut by a transversal so that the exterior angles on the same side of the transversal are supplementary, then these lines are parallel.

Figure 10.11 will help you visualize this situation. Two lines, l and m, are cut by a transversal t, with interior angles on the same side of the transversal labeled $\angle 1$ and $\angle 2$.

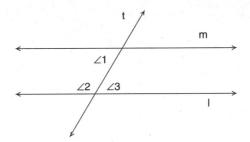

Figure 10.11

l *and* m, *are cut by a transversal* t, *and* ∠1 *and* ∠2 *are interior angles on the same side of the transversal.*

Given: l and m are cut by a transversal t, ∠1 and ∠2 are supplementary angles.

Prove: l‖m.

Proof: Here's the game plan. In order to use Theorem 10.7, you need to show that corresponding angles are congruent. But it might be easier to use Theorem 10.8 if you can show that ∠2 and ∠3 are congruent. You can do that fairly easily, if you apply what you discovered in Example 2 of Chapter 4. Because ∠1 and ∠3 are supplementary angles, and ∠1 and ∠2 are supplementary angles, you can conclude that ∠2 ≅ ∠3 . Then you apply Theorem 10.8 and your work is done.

	Statements	**Reasons**
1.	l and m are two lines cut by a transversal t, ∠1 and ∠2 are supplementary angles.	Given
2.	∠1 and ∠3 are supplementary angles	Definition of supplementary angles
3.	∠2 ≅ ∠3	Example 2, Chapter 4
4.	l‖m	Theorem 10.8

In a complicated world, a complicated theorem requires a complicated drawing. If your drawing is too involved, it could be difficult to decide which lines are parallel because of congruent angles. Consider Figure 10.12. Suppose that ∠1 ≅ ∠3 . Which lines must be parallel? Because ∠1 and ∠3 are corresponding angles when viewing lines o and n cut by transversal m, o‖n.

Figure 10.12

The intersection of lines l, m, n, *and* o.

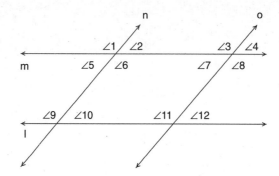

Put Me in, Coach!

Here's your chance to shine. Remember that I am with you in spirit and have provided the answers to these questions in Appendix A.

1. If l || m as in Figure 10.4, with $m\angle 2 = 2x - 45$ and $m\angle 1 = x$, find $m\angle 6$ and $m\angle 8$.

2. Write a formal proof for Theorem 10.3.

3. Write a formal proof for Theorem 10.5.

4. Prove Theorem 10.9.

5. Prove Theorem 10.11.

6. In Figure 10.12, which lines must be parallel if $\angle 3 \cong \angle 11$?

The Least You Need to Know

♦ Given a point A on a line l, there exists a unique line m perpendicular to l which passes through A.

♦ When two parallel lines are cut by a transversal, corresponding, alternate interior and alternate exterior angles are congruent. Same-side interior and exterior angles are supplementary.

♦ If any of the above mentioned angle relationships hold when two lines are cut by a transversal, you can conclude that the two lines are parallel.

Part  Piecing Together Triangles and Quadrilaterals

Triangles are probably the most adored polygons, but they don't let their angles swell because of it (at least in Euclidean geometry). The next four chapters are dedicated to these beloved polygons. You will get to know them inside and out. There are many theorems written about triangles, but the most famous of all is the Pythagorean Theorem. You will have a chance to prove it and use it throughout the rest of the book.

If you add another vertex to a triangle, a quadrilateral is born. Inside every quadrilateral are two triangles waiting to be explored. You will have an opportunity to examine the big names in quadrilaterals: trapezoids, parallelograms, kites, rectangles, squares, and rhombuses. You'll get to write a few proofs and compete in an exciting new game show.

Chapter 11

Two's Company, Three's a Triangle

In This Chapter

◆ Types of triangles

◆ The Pythagorean Theorem

◆ Calculating the perimeter and area of a triangle

◆ The Triangle Inequality

As long as there are three of anything, there will be triangles. You can find three of anything almost everywhere you look. Usually the only triangles you read about in books are love triangles. But in this book there will be no tragedy; no one will be jealous and no one will get hurt.

Triangles have the distinction of being the simplest polygon in that they have the fewest number of sides. You can break up convex polygons into triangles, which is how you were able to calculate the sum of the interior angles of a polygon in Chapter 6. When you think of basic building blocks, you usually think of rectangles, but two triangles can combine to form a rectangle.

Triangles are the shape of choice when designing the wing of a paper airplane, the flap of an envelope, or the side of a pyramid. This chapter gives you the opportunity to get to know them better.

A Formal Introduction

The simplest polygon is a triangle. Given three noncollinear points A, B, and C, triangle ABC is the polygon formed by the sides $\overline{AB}$, $\overline{AC}$, and $\overline{BC}$. Figure 11.1 shows a picture of triangle ABC. The points A, B, and C will be referred to as the *vertices* of the triangle and we will refer to the triangle as $\triangle ABC$. The little triangle before the ABC is there to help you visualize it. A triangle has three sides ($\overline{AB}$, $\overline{AC}$, and $\overline{BC}$) and three angles ($\angle ABC$, $\angle ACB$, and $\angle CAB$). These angles are called the *interior angles* of $\triangle ABC$. Triangles can be classified by either the lengths of their sides or the measures of their interior angles.

Figure 11.1

$\triangle ABC$.

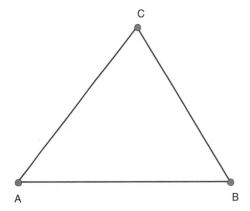

Classifying Triangles by Their Angles or Their Sides

First let's classify triangles according to their size. Remember that two line segments are congruent if they have the same length. If all three sides of a triangle are congruent it is called an *equilateral triangle*. If only two of the sides are congruent, it is called an *isosceles triangle*. If no two sides are congruent, it is a *scalene triangle*. Triangles of each of these types are shown in Figure 11.2.

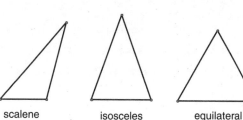

Figure 11.2

Triangles can be classified by the number of congruent sides.

scalene   isosceles   equilateral

Triangles are also classified according to their angles. Recall that the four angle types are acute, right, obtuse, and straight. There is no such thing as a straight triangle, for reasons you will see soon. If a triangle has one obtuse angle, it is an *obtuse triangle*. A triangle cannot have more than one obtuse angle, because the sum of the measures of two obtuse angles is greater than the 180° allocated to a triangle. If a triangle has a right angle, it is a *right triangle*. A triangle cannot have more than one right angle, either. But a triangle can have more than one acute angle. In fact, every triangle has at least two acute angles, and if all three angles of a triangle are acute you have an *acute triangle*. If all three angles of a triangle are congruent, it is an *equiangular triangle*. Triangles of each of these types are shown in Figure 11.3.

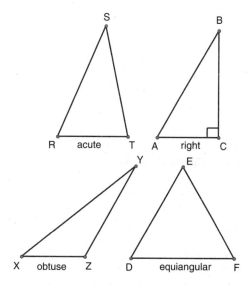

Figure 11.3

Triangles can be classified by angle size.

In a right triangle, the side opposite the right angle is called the hypotenuse of the triangle, and the other two shorter sides are called the legs of the triangle. In the right triangle shown in Figure 11.3, $\overline{AC}$ and $\overline{BC}$ are legs, and $\overline{AB}$ is the hypotenuse of the right triangle $\triangle ABC$.

Solid Facts

Equilateral triangle: All three sides of the triangle are congruent.

Isosceles triangle: At least two sides of the triangle are congruent.

Scalene triangle: No sides of the triangle are congruent.

Obtuse triangle: A triangle with one obtuse angle.

Right triangle: A triangle with one right angle.

Equiangular triangle: A triangle with all angles congruent.

Acute triangle: A triangle with all three angles measuring less than 90°.

In general, a triangle is categorized by both its sides and its angles. For example, a triangle with two congruent sides and a right angle is called an isosceles right triangle. A triangle with one right angle but with no two sides congruent is called a right scalene triangle.

Sums of Interior Angles Are Cooking at 180°

In Chapter 6 you learned how to compute the sum of the interior angles of a polygon using the equation $S = (n - 2) \times 180°$. You used this equation to show that the measures of the interior angles of a triangle add up to 180°. You can use this fact to prove some angle relationships and calculate some angle measures. If you (or anyone else) were to draw a triangle, measure its three angles, and add them together, the sum will always be 180°. It wouldn't matter what kind of triangle you drew, or how big it was. You would always get the same result when you added all three angles together: 180°. I will state this as a theorem, and use it freely from now on.

Theorem 11.1: In a triangle, the sum of the measures of the interior angles is 180°

As a result of this theorem, you can establish two other results.

Suppose $\triangle ABC$ is an equiangular triangle. What can you say about its angles? If all three angles of the triangle are congruent, then all three angles have the same measure. You don't know what the measure is, so you'll call it x. Because the sum of the measures of the interior angles of a triangle is 180°, you know that

$x + x + x = 180°$

$3x = 180°$

$x = 60°$

So, if you have an equiangular triangle, each interior angle measures 60°.

   Theorem 11.2: Each angle of an equiangular triangle measures 60°.

Here are examples of some algebraic problems that can be solved in this geometric environment.

   Example 1: Given that ΔNMQ is a right triangle, and ∠M is a right angle (see Figure 11.4), find m∠Q if m∠N = 44°.

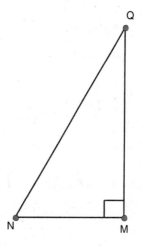

Figure 11.4

ΔNMQ is a right triangle, and ∠M is a right angle.

   Solution: The sum of the interior angles of ΔNMQ is 180°. Thus

$$m\angle M + m\angle N + m\angle Q = 180°.$$

Because ∠M is a right angle, you know that $m\angle M = 90°$. You were given that $m\angle N = 44°$. Substituting into the above equation, you have:

$$90° + 44° + m\angle Q = 180°$$

$$134° + m\angle Q = 180°$$

$$m\angle Q = 46°$$

So $m\angle Q = 46°$. Notice that the measures of ∠N and ∠Q add up to 90°. That means that ∠N and ∠Q are complementary. This is an interesting observation that you can prove for all right triangles.

Solid Facts

   An **exterior angle** of a triangle is the angle formed by a side and an extension of the adjacent side.

Exterior Angle Relationships

When the sides of a triangle are extended, new angles appear on the scene. The angle formed by a side and an extension of the adjacent side is called an *exterior angle* of the triangle. In Figure 11.5, ∠ACD is an exterior angle of ΔABC. The two interior angles, ∠A and ∠B, can be described as nonadjacent interior angles of ∠ACD.

Figure 11.5

∠ACD is an exterior angle of ΔABC.

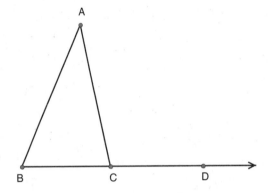

Each vertex of a triangle has two exterior angles, for a grand total of six exterior angles. Figure 11.6 shows all six exterior angles of a triangle.

Figure 11.6

Triangles have six exterior angles.

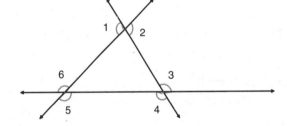

With these new angles, you can work out even more algebraic problems.

Example 2: Suppose ΔMNQ has exterior angle ∠NQR , as in Figure 11.7. If m∠NQR = 117° and m∠M = 50°, find the measures of the other two interior angles of ΔMNQ.

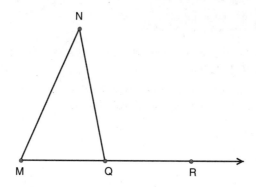

Figure 11.7

$\triangle$MNQ *with exterior angle* $\angle$NQR.

Solution: There are several angle pieces that fit together in this problem. $\angle$MQR is straight, so

$$m\angle MQR = 180°.$$

Using the Angle Addition Postulate, you know that

$$m\angle MQN + m\angle NQR = m\angle MQR.$$

Substituting in for $\angle$MQR and $\angle$NQR, you have

$$m\angle MQN + 117° = 180°$$

$$m\angle MQN = 63°$$

That's one angle down, and one to go. You know the measures of two of the three interior angles, so you can use Theorem 11.1 to find the measure of the third:

$$m\angle M + m\angle N + m\angle NQM = 180°$$

$$m\angle M + m\angle N + m\angle NQM = 180°$$

$$50° + m\angle N + 63° = 180°$$

$$m\angle N = 67°$$

You have found the measure of the other two interior angles: $m\angle N = 67°$ and $m\angle MQN = 63°$.

Notice that $m\angle M + m\angle N = 50° + 67° = 117°$. That's exactly the size of the exterior angle $\angle$NQR. This is not a coincidence. I'll state it as a theorem and let you prove it when you have finished reading this chapter.

Theorem 11.3: The measure of an exterior angle of a triangle equals the sum of the measures of the two nonadjacent interior angles.

Size Matters, So Let's Measure

When I draw a triangle, one of the first things I do after I admire the color scheme and the straightness of the sides is compare my triangle to all other triangles around. Is my triangle the biggest one around? How can I decide what determines the biggest? For example, in Figure 11.8, which of the two triangles is the biggest?

Figure 11.8

Which of these two triangles is bigger?

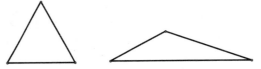

One way to decide which triangle is bigger is to examine which triangle has the longest side. You could also base your decision on which triangle has the largest perimeter. Or you could base everything on the size of the enclosed area. This is a good time to discuss these properties.

Perimeter

Imagine that you have a triangular piece of property that you want to fence off. Before you head off to the lumber store to buy the materials, you will need to know how much fence to buy. You will need to know the lengths of all three sides of the region, and then you can add them together. When you measure the lengths of all three sides of a triangle and add them together, you are measuring the *perimeter* of the triangle.

Solid Facts

The **perimeter** of a triangle is the sum of the lengths of the three sides.

Area

The *area* of a triangle (or any other polygon) is a measurement of the amount of the region within an enclosed plane figure. It is a fairly simple task to calculate the area of a triangle. All that is needed is the length of its base and its height. The area of a triangle can be calculated by evaluating one-half the length of its base times its height.

If you measure the sides of your polygon in feet, then the area of the polygon will have units feet^2 (read square feet). If the sides of our polygon are measured in yards, the units of area will be yards^2 (read square yards). In general, if the length of a side is measured in "units," the units of your area will be units^2 (read square units).

But back to the nuts and bolts of calculating the area of a triangle, I need to explain what I mean by the base and the height. The height of a triangle is related to its altitude. An *altitude* of a triangle is a line segment drawn perpendicularly from a vertex of the triangle to the opposite side of the triangle. The length of the altitude is called the *height* of the triangle. The side opposite the vertex is called the *base* of the triangle.

A triangle's base and altitude depend on how it is oriented. Any side of a triangle can serve as its base. The side that is lowest in our figure is usually referred to as the base. Suppose you are dealing with ΔABC in Figure 11.9. In this case, $\overline{AB}$ would be considered to be the base and $\overline{CD}$ the altitude. The altitude does not have to lie in the interior of the triangle, as is the case with ΔRST, also shown in Figure 11.9.

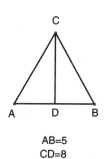
AB=5
CD=8

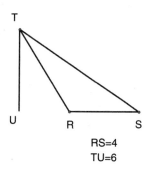
RS=4
TU=6

Figure 11.9

ΔABC *with base* $\overline{AB}$ *and height* CD, *and* ΔRST *with base* $\overline{RS}$ *and altitude* $\overline{TU}$.

Solid Facts

The **area** of a polygon is the measurement of the size of the region within the polygon. The units of area are the units of length squared.

An **altitude** of a triangle is a line segment drawn perpendicularly from a vertex of the triangle to the opposite side of the triangle.

The **height** of a triangle is the length of its altitude.

The **base** of the triangle is the side opposite the vertex.

It is possible to calculate the area of ΔABC in Figure 11.9. All that has to be done is evaluate one-half the product of the length of its base and its height.

Example 3: Find the area of ΔABC.

Solution: The base of ΔABC has length 5, and its height is 8. The area of ΔABC is $\frac{1}{2} \times 5 \times 8 = 20$.

> **Eureka!** _____
>
> The area of a right triangle can be calculated by taking one-half the product of the lengths of the two legs.

Computing the area of a right triangle is fairly easy, if it has the right orientation. I recommend always using one leg for the base. Then, because the legs of a right triangle are perpendicular to each other, the other leg will serve as the altitude. So the area of a right triangle is one-half the product of the lengths of the legs. Figure 11.10 will shed some light on this situation.

Figure 11.10

$\triangle ABC$ _has area_

½(AC) × (BC).

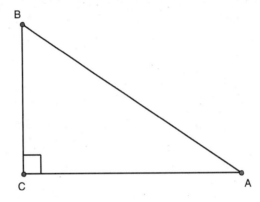

Ladies and Gentlemen: The Pythagorean Theorem

The Pythagorean Theorem is one of the best known theorems in the world. There are several slick ways to prove this theorem. One method involves similar triangles (which you will learn about in Chapter 13). Another technique makes use of the area of squares, which you will examine in Chapter 15. As a theorem it is so useful that I can't wait to share it with you. Accept it as true for now and practice using it.

> **Theorem 11.4:** The Pythagorean Theorem. The square of the length of the hypotenuse of a right triangle is equal to the sum of the squares of the lengths of the legs.

Practice using this theorem right now!

> **Example 4:** Given the right triangle shown in Figure 11.11, with AB = 5, and BC = 3, find the area of $\triangle ABC$.

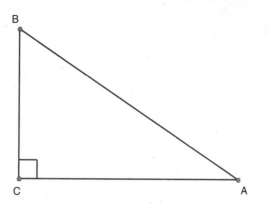

Figure 11.11
$\triangle$ABC.

Solution: You are given that the hypotenuse of $\triangle$ABC has length 5, and that the length of one of the legs is 3. In order to find the area of a right triangle, all you need to know are the lengths of the two legs. You are given the length of one leg and the hypotenuse. You need the length of the second leg. The Pythagorean Theorem is here to save the day.

$$(AC)^2 + (BC)^2 = (AB)^2$$
$$(AC)^2 + 3^2 = 5^2$$
$$(AC)^2 = 25 - 9 = 16$$
$$AC = 4$$

Now that you know the length of both legs, you can find the area:

Area = $\frac{1}{2}$(AC) $\times$ (BC) = $\frac{1}{2}$(4)(3) = 6. So the area of $\triangle$ABC is 6.

The Triangle Inequality

Sometimes things in life are just not equal. Your neighbor might have more money than you, or you might be smarter than some of your friends. As much as mathematicians would like everything to be equal in the world, it can't be. Inequality exists, and there's nothing that can be done about it. Some numbers are bigger than others. And some aspects of geometric figures are unequal as well. The Triangle Inequality is a theorem about the inequity of the sides of a triangle.

> **Theorem 11.5:** Triangle Inequality. The sum of the lengths of any two sides of a triangle is greater than the length of the third side.

You can write a formal proof of the triangle inequality. Figure 11.12 shows $\triangle$ABC.

Figure 11.12

$\triangle ABC$.

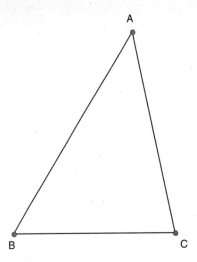

Given: $\triangle ABC$

Prove: $AB + AC > BC$

Proof: You need a game plan. If you drop a perpendicular segment from A to $\overline{AB}$, you can use the Pythagorean Theorem to show that $AB > BD$ and $AC > DC$ (see Figure 11.13). Whenever you add some detail to your drawing, such as introducing perpendicular lines, your justification is "construction." You are allowed to construct things like parallel and perpendicular lines by Euclid's Parallel Postulate and Theorem 10.1. The Additive Property of Inequality shows that $AB + AC > BD + DC$. If you apply the Segment Addition Postulate to show that $BD + DC = BC$, you are done. Bring out the columns!

Figure 11.13

$\triangle ABC$ *with* $\overline{AD} \perp \overline{BC}$.

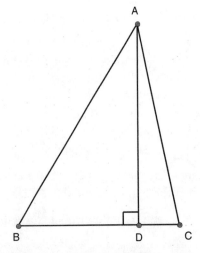

	Statements	Reasons
1.	$\triangle ABC$	Given
2.	Draw $\overline{AD} \perp \overline{BC}$	Construction (or Theorem 10.1)
3.	$(BD)^2 + (AD)^2 = (AB)^2$ and $(CD)^2 + (AD)^2 = (AC)^2$	The Pythagorean Theorem
4.	$(AB)^2 > (BD)^2$ and $(AC)^2 > (DC)^2$	Definition of $>$
5.	$AB > BD$ and $AC > DC$	Algebra
6.	$AB + AC > BD + DC$	Additive Property of Inequality
7.	$BD + DC = BC$	Segment Addition Postulate
8.	$AB + AC > BC$	Substitution (steps 6 and 7)

You can use the Triangle Inequality to determine if triangles with certain dimensions can exist.

> **Example 5**: Can a triangle have sides with length 3, 4, and 7?

> **Solution**: By the Triangle Inequality, the length of any one side must be less than the sum of the lengths of the other two sides. Because 3 + 4 = 7, this triangle violates the Triangle Inequality and cannot exist.

Put Me in, Coach!

Here's your chance to shine. Remember that I am with you in spirit and have provided the answers to these questions in Appendix A.

1. What would you call a triangle with two sides congruent and one obtuse angle?

2. Prove that the acute angles of a right triangle are complementary.

3. Write an informal (two-column proof) of Theorem 11.3.

4. Find the area of $\triangle RST$ in Figure 11.9.

5. Find the area of $\triangle RST$ shown in Figure 11.14, if RS = 13 and ST = 5.

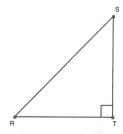

Figure 11.14

$\triangle RST$.

6. Can a triangle have sides of length 3, 4, and 8?

The Least You Need to Know

♦ The measures of the interior angles of a triangle add up to 180°.

♦ The area of a triangle is one-half the base times the height.

♦ The Pythagorean Theorem states that the square of the length of the hypotenuse of a right triangle is equal to the sum of the squares of the lengths of the legs.

♦ The Triangle Inequality theorem states that the sum of the lengths of any two sides of a triangle is greater than the length of the third side.

Chapter 12

Congruent Triangles

In This Chapter

- ◆ CPOCTAC
- ◆ SSS, SAS, and ASA Postulates
- ◆ AAS and HL Theorems
- ◆ More proofs!

It's not enough that mathematicians explore their own individual triangles. Once we become comfortable with triangle ownership, it's in our nature to compare our triangle to all the other triangles out there. We need to know when two triangles match.

Every triangle has six parts (three angles and three sides). So the only way that two triangles can match is if they line up angle for angle, and side for side. The flip side is that if two triangles line up angle for angle, side for side, then they match.

With attention spans being what they are, most of us don't have the patience required to match up all six parts of a triangle to all six parts of another triangle. Fortunately for us, there is an easier way.

CPOCTAC

Two triangles are *congruent* if one fits perfectly over the other. It should come as no surprise that you have to be specific about exactly how these two triangles fit together. After all, there are three angles, and three sides, and you wouldn't want to pair wrong angles or the wrong sides together. Figure 12.1 shows two triangles: $\triangle ABC$ and $\triangle DEF$. These two triangles look like identical twins. You can say that $\triangle ABC$ and $\triangle DEF$ are congruent (and write $\triangle ABC \cong \triangle DEF$) if $\angle A \cong \angle D$, $\angle B \cong \angle E$, $\angle C \cong \angle F$, and $\overline{AB} \cong \overline{DE}$, $\overline{AC} \cong \overline{DF}$, and $\overline{BC} \cong \overline{EF}$. In this situation, vertex A corresponds to vertex D, vertex B corresponds to vertex E, and vertex C corresponds to vertex F.

Figure 12.1

$\triangle ABC$ *and* $\triangle DEF$ *are congruent triangles.*

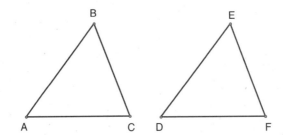

The order that you write the vertices of the triangle shows which angles and which sides are congruent. You can also talk about corresponding angles: $\angle A$ and $\angle D$ are corresponding angles, as are $\angle B$ and $\angle E$, and $\angle C$ and $\angle F$. This definition of congruence works both ways. If you are told that $\triangle ABC \cong \triangle MNO$, then right away you know that their corresponding parts are congruent: for example, $\angle A \cong \angle M$ and so on. The notion that *corresponding parts of congruent triangles are congruent* will be used so often that it will be abbreviated CPOCTAC. It will play a crucial role in proving the congruence of line segments and angles.

Solid Facts

Two triangles are **congruent** when the six parts of the first triangle are congruent to the six corresponding parts of the second triangle.

As you have already seen with congruent line segments and congruent angles, the notion of congruence with respect to triangles is an equivalence relation. It has the reflexive, symmetric, and transitive properties, which are easy to verify.

The Big Five

In order to make use of the fact that corresponding parts of congruent triangles are congruent you will need to be able to prove that the two triangles are, in fact, congruent without using the definition of triangle congruence. I'll give you some shortcuts (in the form of postulates and theorems) that will help you do this. Most of these postulates and theorems will be known by their abbreviations: S stands for side and A stands for angle.

Eureka! _____

To come up with a game plan for proving triangles congruent, here is some helpful advice:

1. Mark the figures systematically. Use a square in the opening of a right angle, use the same number of dashes on congruent sides and use the same number of arcs on congruent angles.

2. Trace the triangles suspected to be congruent in different colors.

3. If the triangles overlap, draw them separately.

The SSS Postulate

As the name implies, you can conclude that two triangles are congruent based on just the lengths of the sides of two triangles.

Postulate 12.1: SSS Postulate. If the three sides of one triangle are congruent to the three sides of a second triangle, then the triangles are congruent.

Let's practice using this postulate. Take a look at Figure 12.2. Suppose that $\overline{AB}$ and $\overline{CD}$ bisect each other at M and that $\overline{AC} \cong \overline{DB}$. Prove that $\triangle AMC \cong \triangle BMD$.

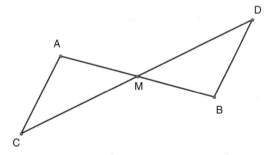

Figure 12.2

$\overline{AB}$ *and* $\overline{CD}$ *bisect each other at* M, *and* $\overline{AC} \cong \overline{DB}$.

Example 1: Given the situation shown in Figure 12.2, if $\overline{AB}$ and $\overline{CD}$ bisect each other at M, and $\overline{AC} \cong \overline{DB}$, write an informal proof that $\triangle AMC \cong \triangle BMD$.

Solution: You can skip the formalities of rewriting the statement you are trying to prove. You already have a picture (Figure 12.2) to help you visualize the situation. You are given that $\overline{AB}$ and $\overline{CD}$ bisect each other at M, and $\overline{AC} \cong \overline{DB}$, and you need to show that $\triangle AMC \cong \triangle BMD$. To make use of the SSS Postulate you will need to explore the relationships between the sides of the two triangles. Because $\overline{AB}$ and $\overline{CD}$ bisect each other at M you know that $\overline{AM} \cong \overline{MB}$ and $\overline{CM} \cong \overline{MD}$. That should be enough to nail down the details of the proof.

	Statements	**Reasons**
1.	$\overline{AB}$ and $\overline{CD}$ bisect each other at M, and $\overline{AC} \cong \overline{DB}$	Given
2.	$\overline{AM} \cong \overline{MB}$ and $\overline{CM} \cong \overline{MD}$	Definition of segment bisector
3.	$\triangle AMC \cong \triangle BMD$	SSS Postulate

The SAS Postulate

You are probably familiar with what it means to be included: You are a part of the group; you belong. Angles and line segments aren't much different. Line segments can include an angle, and angles can include a line segment. The two sides of a triangle that form an angle are said to *include that angle* of the triangle. Similarly, any two angles of a triangle must have a common side, and these two angles are said to *include that side*. For example, Figure 12.3 shows a picture of $\triangle ABC$. $\overline{AC}$ and $\overline{CB}$ include $\angle C$, and $\angle A$ and $\angle B$ include $\overline{AB}$.

Figure 12.3

In $\triangle ABC$, $\overline{AC}$ and $\overline{CB}$ include $\angle C$, and $\angle A$ and $\angle B$ include $\overline{AB}$.

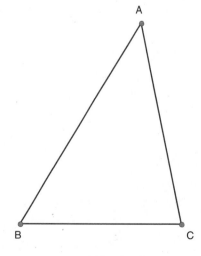

You can use included angles and line segments to prove that two triangles are congruent.

> **Postulate 12.2:** SAS Postulate. If two sides and the included angle of one triangle are congruent to two sides and the included angle of a second triangle, then the triangles are congruent.

The order of the letters in the name SAS Postulate will help you remember that the two sides that are named actually form the angle.

> **Example 2**: If $\overline{PN} \perp \overline{MQ}$ and $\overline{MN} \cong \overline{NQ}$ as shown in Figure 12.4, write a two-column proof that $\triangle PNM \cong \triangle PNQ$.

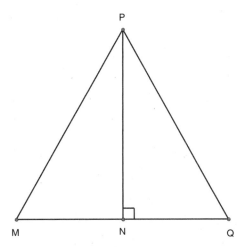

Figure 12.4
$\overline{PN} \perp \overline{MQ}$ *and* $\overline{MN} \cong \overline{NQ}$.

Solution: The game plan is to make use of the SAS Postulate. Because $\overline{PN} \perp \overline{MQ}$, you know that right angles are formed. And two right angles are congruent. The congruence of one set of sides is given. Use the reflexive property of $\cong$ to obtain another set of congruent sides: A side is congruent to itself. Notice that the angles you are focusing on are $\angle MNP$ and $\angle QNP$. The sides that include these angles are $\overline{MN}$ and $\overline{PN}$ (for $\angle MNP$) and $\overline{NQ}$ and $\overline{PN}$ (for $\angle QNP$). As long as you are careful to discuss two sides and the included angle, you'll be fine.

	Statements	Reasons
1.	$\overline{PN} \perp \overline{MQ}$ and $\overline{MN} \cong \overline{NQ}$	Given
2.	$\angle MNP$ and $\angle QNP$ are right angles	Definition of $\perp$
3.	$m\angle MNP = 90°$ and $m\angle QNP = 90°$	Definition of right angle
4.	$m\angle MNP = m\angle QNP$	Substitution
5.	$\angle MNP \cong \angle QNP$	Definition of $\cong$

	Statements	Reasons
6.	$\overline{PN} \cong \overline{PN}$	Reflexive property of $\cong$ ·
7.	$\triangle PNM \cong \triangle PNQ$	SAS Postulate

You will see several theorems about isosceles triangles. Most of these will be proven using the SAS postulate. For example, if $\triangle ABC$ is an isosceles triangle with $\overline{AB} \cong \overline{BC}$, you can show that $\triangle ABC \cong \triangle CBA$ by SAS. Thus $\angle A \cong \angle C$ by CPOCTAC. These are the angles opposite the congruent sides in $\triangle ABC$. This is the first of many theorems about isosceles triangles.

>**Theorem 12.1:** In an isosceles triangle, the angles opposite the congruent sides are congruent.

The ASA Postulate

You can prove that two triangles are congruent if you know about all three sides of the triangles, or if you know about two sides and the included angle. This next postulate will enable you to prove that two triangles are congruent based on two angles and the included side.

>**Postulate 12.3:** The ASA Postulate. If two angles and the included side of one triangle are congruent to two angles and the included side of a second triangle, then the triangles are congruent.

Let's see how this postulate can be used.

Example 3: If $\overline{AC} \cong \overline{DC}$ and $\angle A \cong \angle D$, as shown in Figure 12.5, write a two-column proof to show that $\triangle ACE \cong \triangle DCB$.

Figure 12.5

$\overline{AC} \cong \overline{DC}$ *and* $\angle A \cong \angle D$.

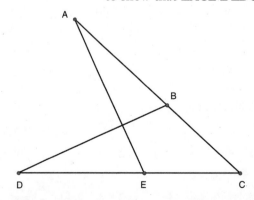

Solution: The figures are starting to get complicated. Whenever this happens, break things apart to get a grip on each piece of the puzzle. Figure 12.6 might help you sort through the mess.

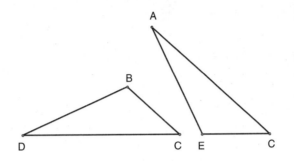

Figure 12.6

The triangles in Figure 12.5 pulled apart. You still have that $\overline{AC} \cong \overline{DC}$ *and* $\angle A \cong \angle D$.

Notice that $\angle C \cong \angle C$ (by the reflexive property of $\cong$). Because you are given that $\angle A \cong \angle D$, you have information about two angles. You just need to know about the included sides. The included side for $\triangle ACE$ is $\overline{AC}$, and the included side for $\triangle DCB$ is $\overline{DC}$. You are given that those two sides are congruent! That enables you to use the ASA Postulate! Write out the details.

Statements	Reasons
1. $\overline{AC} \cong \overline{DC}$ and $\angle A \cong \angle D$	Given
2. $\angle C \cong \angle C$	Reflexive property of $\cong$
3. $\triangle ACE \cong \triangle DCB$	ASA Postulate

Notice that the proof is fairly short. The hard part is pulling the picture apart to see how the pieces relate. That's not unusual. The hardest part about writing a proof usually involves coming up with the game plan.

The AAS Theorem

You've accepted several postulates in this chapter. That's enough faith for a while. It's time for your first theorem, which will come in handy when trying to establish the congruence of two triangles.

> **Theorem 12.2:** The AAS Theorem. If two angles and a nonincluded side of one triangle are congruent to two angles and a nonincluded side of a second triangle, then the triangles are congruent.

Figure 12.7 will help you visualize the situation. In the following formal proof, you will relate two angles and a nonincluded side of $\triangle ABC$ to two angles and a nonincluded side of $\triangle RST$.

Figure 12.7

Two angles and a nonin-
cluded side of △ABC are
congruent to two angles and
a nonincluded side of △RST .

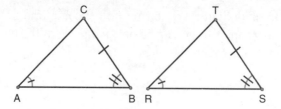

Given: Two triangles, △ABC and △RST, with ∠A ≅ ∠R , ∠C ≅ ∠T, and BC ≅ ST.

Prove: △ABC ≅ △RST.

Proof: You need a game plan. If only you knew about two angles and the included side! Then you would be able to use the ASA Postulate to conclude that △ABC ≅ △RST. But wait a minute! Because the measures of the interior angles of a triangle add up to 180°, and you know two of the angles in △ABC are congruent to two of the angles in △RST, you can show that the third angle of △ABC is congruent to the third angle in △RST. Then you'll have two angles and the included side of △ABC congruent to two angles and the included side of △RST, and you're home free.

	Statements	Reasons
1.	△ABC and △RST with ∠A ≅ ∠R , ∠C ≅ ∠T , and BC ≅ ST.	Given
2.	m∠A = m∠R and m∠C = m∠T	Definition of ≅
3.	m∠A + m∠B + m∠C = 180° and m∠R + m∠S + m∠T = 180°	Theorem 11.1
4.	m∠A + m∠B + m∠C = m∠R + m∠S + m∠T	Substitution (step 3)
5.	m∠A + m∠B + m∠C = m∠A + m∠S + m∠C	Substitution (steps 2 and 4)
6.	m∠B = m∠S	Algebra
7.	∠B ≅ ∠S	Definition of ≅
8.	△ABC ≅ △RST	ASA Postulate

The HL Theorem for Right Triangles

Whenever you are given a right triangle, you have lots of tools to use to pick out important information. For example, not only do you know that one of the angles of the triangle is a right angle, but you know that the other two angles must be acute angles. You also have the Pythagorean Theorem that you can apply at will. Finally, you know that the two legs of the triangle are perpendicular to each other. You've made use of the perpendicularity of the legs in the last two proofs you wrote on your own. Now it's time to make use of the Pythagorean Theorem.

> **Theorem 12.3:** The HL Theorem for Right Triangles. If the hypotenuse and a leg of one right triangle are congruent to the hypotenuse and a leg of a second right triangle, then the triangles are congruent.

There are several ways to prove this problem, but none of them involve using an SSA Theorem. Your plate is so full with initialized theorems that you're out of room. Not to mention the fact that a SSA relationship between two triangles is not enough to guarantee that they are congruent. If you use the Pythagorean Theorem, you can show that the other legs of the right triangles must also be congruent. Then it's just a matter of using the SSS Postulate.

Figure 12.8 illustrates this situation. You have two right triangles, $\triangle ABC$ and $\triangle RST$.

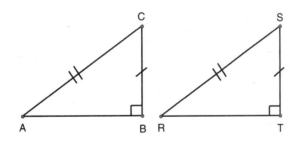

Figure 12.8

The hypotenuse and a leg of $\triangle ABC$ are congruent to the hypotenuse and a leg of $\triangle RST$.

Given: $\triangle ABC$ and $\triangle RST$ are right triangles with $\overline{AB} \cong \overline{RS}$ and $\overline{BC} \cong \overline{ST}$.

Prove: $\triangle ABC \cong \triangle RST$.

Proof: You already have a game plan, so all that's left is to execute it.

Statements	Reasons
1. $\triangle ABC$ and $\triangle RST$ are right triangles with $\overline{AB} \cong \overline{RS}$ and $\overline{BC} \cong \overline{ST}$.	Given
2. $AB = RS$ and $BC = ST$	Definition of $\cong$
3. $(AC)^2 + (BC)^2 = (AB)^2$ and $(RT)^2 + (ST)^2 = (RS)^2$	The Pythagorean Theorem
4. $(AC)^2 + (BC)^2 = (RT)^2 + (ST)^2$	Substitution (steps 2 and 3)
5. $(AC)^2 + (ST)^2 = (RT)^2 + (ST)^2$	Substitution (steps 2 and 4)
6. $(AC)^2 = (RT)^2$	Algebra
7. $AC = RT$	Algebra
8. $\overline{AC} \cong \overline{RT}$	Definition of $\cong$
9. $\triangle ABC \cong \triangle RST$	SSS Postulate

 Tangled Knot _____

SSS, SAS, ASA, and AAS are valid methods of proving triangles congruent, but SSA and AAA are **not** valid methods and cannot be used. In Figure 12.9, the two triangles are marked to show SSA, yet the two triangles are not congruent. Figure 12.10 shows two triangles marked AAA, but these two triangles are also not congruent.

Figure 12.9

These two triangles are not congruent, even though two corresponding sides and an angle are congruent. The two congruent sides do not include the congruent angle!

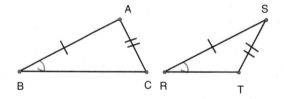

Figure 12.10

These two triangles are not congruent, even though all three corresponding angles are congruent.

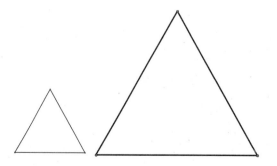

Proving Segments and Angles Are Congruent

After you have shown that two triangles are congruent, you can use the fact that CPOCTAC to establish that two line segments (corresponding sides) or two angles (corresponding angles) are congruent.

> **Example 4**: If ∠R and ∠V are right angles, and ∠RST ≅ ∠VST (see Figure 12.11), write a two-column proof to show $\overline{RT} \cong \overline{TV}$.

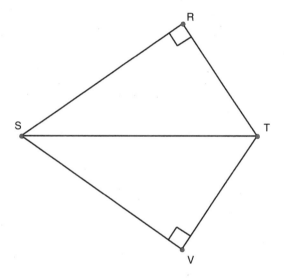

Figure 12.11

∠R and ∠V are right angles, and ∠RST ≅ ∠VST .

> **Solution**: You need a game plan. If you could show that ΔRST ≅ ΔVST, then you could use CPOCTAC to show that $\overline{RT} \cong \overline{TV}$. To show that ΔRST ≅ ΔVST, you simply use the AAS Theorem.

	Statements	Reasons
1.	∠R and ∠V are right angles, and ∠RST ≅ ∠VST	Given
2.	ΔRST ≅ ΔVST	AAS Theorem
3.	$\overline{RT} \cong \overline{TV}$	CPOCTAC

> **Example 5**: Suppose that in Figure 12.12, $\overrightarrow{CB}$ bisects ∠ACD and $\overline{BC} \perp \overline{AD}$. Write a two-column proof to show that ∠A ≅ ∠D .

Figure 12.12

$\overrightarrow{CB}$ *bisects* ∠ACD *and*
$\overline{BC} \perp \overline{AD}$.

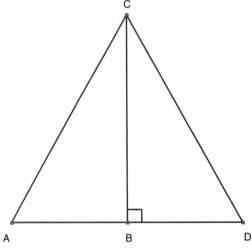

Solution: Because $\overline{BC} \perp \overline{AD}$, you know that ∠ABC ≅ ∠DBC. Because $\overrightarrow{CB}$ bisects ∠ACD , you know that ∠ACB ≅ ∠DCB. Finally, $\overline{BC}$ is congruent to itself, and you can use the ASA Postulate to show that △ABC ≅ △DBC. By CPOCTAC, you can conclude that ∠A ≅ ∠D. Let's write it up.

Statements	Reasons
1. $\overrightarrow{CB}$ bisects ∠ACD and $\overline{BC} \perp \overline{AD}$	Given
2. ∠ABC and ∠DBC are right angles	Definition of ⊥
3. m∠ABC = 90° and m∠DBC = 90°	Definition of right angle
4. m∠ABC = m∠DBC	Substitution
5. ∠ABC ≅ ∠DBC	Definition of ≅
6. ∠ACB ≅ ∠DCB	Definition of angle bisector
7. $\overline{BC} \cong \overline{BC}$	Reflexive property of ≅
8. △ABC ≅ △DBC	ASA Postulate
9. ∠A ≅ ∠D	CPOCTAC

Proving Lines Are Parallel

CPOCTAC can be used to prove that line segments are congruent. It can also be used to prove that line segments are parallel. Recall that you have a couple of theorems to help you conclude that two lines are parallel. In Theorem 10.8 you learned that if two lines are cut by a transversal so that alternate interior angles are congruent, then those two lines are parallel. You will use this theorem in conjunction with CPOCTAC to show that line segments are parallel.

Example 6: Suppose that $\overline{AC} \cong \overline{BD}$ and $\overline{AB} \cong \overline{CD}$, as shown in Figure 12.13. Write a two-column proof to show that $\overline{AB} \| \overline{CD}$.

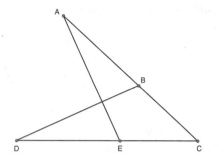

Figure 12.13
$\overline{AC} \cong \overline{BD}$ *and* $\overline{AB} \cong \overline{CD}$.

Solution: Your game plan is to involve showing that $\triangle ABC \cong \triangle DCB$. You have to be careful about how you write these triangles. It's easy to see that you will want $\angle A$ and $\angle D$ to correspond. It's a bit more difficult to see which other vertices need to be paired up. At this point, it would be wise to break the two triangles apart so that the relationships will be clearer. Check out Figure 12.14 to get a better idea of what is going on.

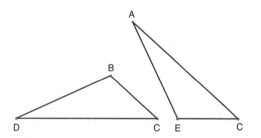

Figure 12.14

The two triangles in Figure 12.13 broken apart.

Because $\overline{BC}$ is congruent to itself, you can use the SSS Postulate to show that $\triangle ABC \cong \triangle DCB$. You can then visualize Figure 12.18 as two lines, $\overleftrightarrow{AB}$ and $\overleftrightarrow{CD}$, cut by a transversal $\overleftrightarrow{BC}$. Then not only are $\angle ACB$ and $\angle DBC$ corresponding angles (and therefore congruent) but they are also alternate interior angles. Because the alternate interior angles are congruent, you know that $\overline{AB} \| \overline{CD}$. You might want to read this paragraph again just to get a good feel for what you are about to do.

Statements	Reasons
1. $\overline{AC} \cong \overline{BD}$ and $\overline{AB} \cong \overline{CD}$	Given
2. $\overline{BC} \cong \overline{BC}$	Reflexive property of $\cong$
3. $\triangle ABC \cong \triangle DCB$	SSS Postulate
4. $\angle ACB \cong \angle DBC$	CPOCTAC

Statements	Reasons
5. $\overleftrightarrow{AB}$ and $\overleftrightarrow{CD}$ are cut by a transversal $\overleftrightarrow{BC}$	Definition of transversal
6. ∠ACB and ∠DBC are alternate interior angles	Definition of alternate interior angles
7. $\overline{AB} \| \overline{CD}$	Theorem 10.8

Put Me in, Coach!

Here's your chance to shine. Remember that I am with you in spirit and have provided the answers to these questions in Appendix A.

1. Verify that ≅ when applied to triangles is an equivalence relation.

2. If $\overline{AC} \cong \overline{CD}$ and ∠ACB ≅ ∠DCB as shown in Figure 12.15, prove that ΔACB ≅ ΔDCB .

Figure 12.15

$\overline{AC} \cong \overline{CD}$ and

∠ACB ≅ ∠DCB .

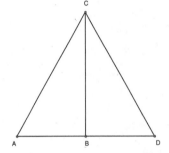

3. If $\overline{CB} \perp \overline{AD}$ and ∠ACB ≅ ∠DCB , as shown in Figure 12.16, prove that ΔACB ≅ ΔDCB .

Figure 12.16

$\overline{CB} \perp \overline{AD}$ and

∠ACB ≅ ∠DCB.

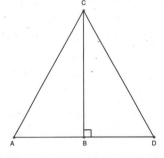

4. If $\overline{CB} \perp \overline{AD}$ and $\angle CAB \cong \angle CDB$, as shown in Figure 12.17, prove that $\triangle ACB \cong \triangle DCB$.

Figure 12.17

$\overline{CB} \perp \overline{AD}$ and

$\angle CAB \cong \angle CDB$.

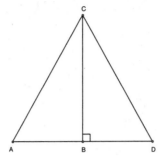

5. If $\overline{CB} \perp \overline{AD}$ and $\overline{AC} \cong \overline{CD}$, as shown in Figure 12.18, prove that $\triangle ACB \cong \triangle DCB$.

Figure 12.18

$\overline{CB} \perp \overline{AD}$ and $\overline{AC} \cong \overline{CD}$.

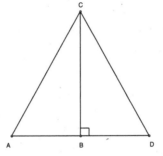

6. If $\angle P \cong \angle R$ and M is the midpoint of $\overline{PR}$, as shown in Figure 12.19, prove that $\angle N \cong \angle Q$.

Figure 12.19

$\angle P \cong \angle R$ *and* M *is the midpoint of* $\overline{PR}$.

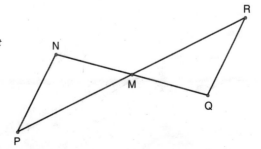

The Least You Need to Know

- There are three postulates used to prove that triangles are congruent: the SSS Postulate, the SAS Postulate, and the ASA Postulate.

- There are two theorems used to prove that triangles are congruent: the AAS Theorem and the HL Theorem for Right Triangles.

- Corresponding parts of congruent triangles are congruent (CPOCTAC).

Similar Triangles

In This Chapter

- ◆ Means-Extremes property
- ◆ AAA Similarity Postulate
- ◆ AA, SAS, and SSS Similarity Theorems
- ◆ CSSTAP

When you have two congruent triangles, you have triangles that are exact duplicates. If you put one on top of the other (with the correct orientation), you wouldn't be able to see the one on the bottom. If all of the triangles in the world were congruent, all paper airplanes would look alike.

Sometimes your triangles have the same shape but not the same size. You might think that all triangles have the same shape, because they are the same kind of polygon. But that's not what I mean when I am talking about triangles having the same shape. I mean that one triangle is an enlargement of the other (or from the other perspective, one triangle is the reduction of the other). The key to dealing with similar triangles is to put things in proportion.

Ratio, Proportion, and Geometric Means

Before you start analyzing similar triangles, you need to pick up a few more algebraic supplies. You'll be dealing with fractions and ratios a lot in this chapter, so you might as well brush up on those algebra skills first.

A ratio is a quotient $\frac{a}{b}$, where $b \neq 0$. A ratio provides a comparison between the numbers a and b. For example, if a is twice as big as b, then the ratio $\frac{a}{b}$ is $\frac{2}{1}$. The ratio $\frac{a}{b}$ is read "a to b" and is sometimes written in the form a:b.

Eureka!

An extended ratio compares more than two quantities and cannot be expressed as a single fraction.

There are times when you might want to compare three or more items. When that happens, a simple fraction just won't cut it. You'll need to use what is called an extended ratio. An extended ratio is written in the form a:b:c (if you are comparing three quantities) or a:b:c:d (if you are comparing four quantities). If you are comparing lots of quantities, just keep adding them on, separating each quantity with a colon.

A proportion is a statement that two ratios are equal. The proportion $\frac{a}{b} = \frac{c}{d}$ is read as "a is to b as c is to d." The first and last terms (a and d) are called the *extremes*, and the middle terms (b and c) are called the *means*. There are several useful properties involving proportions, and these properties can be established using algebra.

> **Property 1:** The Means-Extremes Property. In a proportion, the product of the means equals the product of the extremes. That is, if $\frac{a}{b} = \frac{c}{d}$ (where $b \neq 0$ and $d \neq 0$), then $a \cdot d = b \cdot c$.

This is just an old-fashioned "cross-multiply" step used in algebra to get rid of unwanted denominators when dealing with fractions. You can use the Means-Extremes Property to solve algebraic equations.

> **Example 1:** Use the Means-Extremes Property to solve for x: $\frac{x+1}{9} = \frac{x-3}{3}$.
>
> **Solution:** If we apply the Means-Extremes Property to our equation, we have

$3(x + 1) = 9(x - 3)$

$3x + 3 = 9x - 27$

$6x = 30$

$x = 5$

Example 2: Use the Means-Extremes Property to solve for x: $\dfrac{4}{x} = \dfrac{x}{9}$.

Solution: If you apply the Means-Extremes Property to your equation, you have

$36 = x^2$

$x = \pm\sqrt{36} = \pm 6$.

Now combine these ideas with a problem involving geometry.

Example 3: Suppose that two complementary angles are in the ratio 2 to 3. Find the measure of each angle.

Solution: Let one of your angles have measure x and the other angle have measure y. Because your two angles are complementary, x + y = 90°, so y = 90° – x. Because the ratio of the angle measurements is 2 to 3, you have the following proportion:

$$\dfrac{x}{y} = \dfrac{x}{90° - x} = \dfrac{2}{3}$$

You can then use the Means-Extremes Property:

$3x = 2(90° - x)$

$3x = 180° - 2x$

$5x = 180°$

$x = 36°$

Now that you know the measure of one of your angles, you can find the measure of the second angle because the two angles are complementary:

$y = 90° - x = 90° - 36° = 54°$

There are other properties of proportions. You can flip the proportions and mix and match numerators and denominators. Property 2 of proportions provides a list of the changes you can make.

Property 2: In a proportion, the means or the extremes (or both the means and the extremes) may be interchanged. That is, if $\dfrac{a}{b} = \dfrac{c}{d}$ and a, b, c, and d are all non-zero, then $\dfrac{a}{c} = \dfrac{b}{d}$, $\dfrac{d}{b} = \dfrac{c}{a}$ and $\dfrac{d}{c} = \dfrac{b}{a}$.

Again, mixing numerators and denominators is not surprising because you've been doing this in algebra for years! But there is another property of proportions that might be a bit surprising. You have to be careful when you apply this property because it involves adding or subtracting things.

Property 3: If $\dfrac{a}{b} = \dfrac{c}{d}$, where $b \neq 0$ and $d \neq 0$, then $\dfrac{a+b}{b} = \dfrac{c+d}{d}$.

To derive this property, start with $\dfrac{a}{b} = \dfrac{c}{d}$, add 1 to both sides and add the fractions.

Now that you have proportions under your belt, you can talk about the geometric mean of two numbers. If you start with a proportion where the two means are identical (and the two extremes may be different), such as $\dfrac{a}{b} = \dfrac{b}{d}$, then b is the geometric mean of a and d. You found the geometric means of the numbers 4 and 9 in Example 2. Although a pair of numbers actually has two geometric means (one positive and the other negative), geometers are only interested in the positive one. After one more example you'll be ready to apply these algebraic properties to geometry.

> **Example 4**: In Figure 13.1, suppose that AD is the geometric mean of BD and DC. If BC = 13 and BD = 9, find AD.

Figure 13.1

AD *is the geometric mean of*
BD *and* DC, BC = 13 *and*
BD = 9.

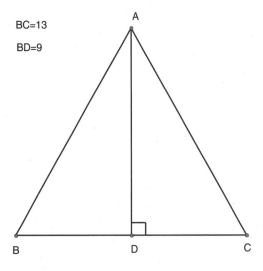

Solution: Because AD is the geometric mean of BD and DC, You know that $\dfrac{BD}{AD} = \dfrac{AD}{DC}$. You are given that BD = 9, but you still need to find DC. Using the Segment Addition Postulate, BC = BD + DC. Substituting in for BC and BD gives 13 = 9 + DC, or DC = 4. Now you can substitute the values into the proportion and solve for AD:

$$\frac{BD}{AD} = \frac{AD}{DC}$$

$$\frac{9}{AD} = \frac{AD}{4}$$

$$(AD)^2 = 36$$

$$AD = 6$$

Properties of Similar Triangles

When we talk about two things being similar, we are trying to convey that our two objects are a lot alike. They don't have to be interchangeable or identical, but they do have to have enough in common. You might think that all triangles are similar, because they have the same number of sides and the same number of angles. But similarity is a special relationship between only certain triangles. In order for two triangles to be declared similar, they must satisfy certain angle and length criteria.

Two triangles are *similar* if all pairs of corresponding angles are congruent and all pairs of corresponding sides are proportional. If we look at the two triangles in Figure 13.2, we can specify the corresponding angles and the various proportionalities that must be satisfied.

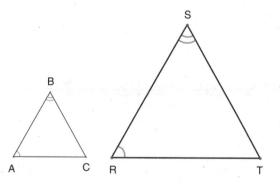

Figure 13.2

Two similar triangles.

The angle congruencies are: $\angle A \cong \angle R$, $\angle B \cong \angle S$ and $\angle C \cong \angle T$. The proportionalities involved are: $\dfrac{AB}{RS} = \dfrac{BC}{ST} = \dfrac{AC}{RT}$ and all possible rearrangements using the various properties of proportions.

I will use the symbol ~ to indicate that two triangles are similar. If you run through the list, you will see that ~ is an equivalence relation, so it will have reflexive, symmetric, and transitive properties.

If you know that two triangles are similar, you can use the extended proportionalities to learn more about the triangles.

> **Example 5**: If ΔABC ~ ΔRST as shown in Figure 13.3, use the indicated measures to find the measures of the remaining sides and angles of each of the triangles.

Figure 13.3

ΔABC ~ ΔRST.

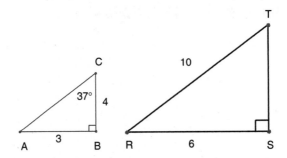

Solution: You have quite a list of things to find: the measure of the third angle in ΔABC; all three angles of ΔRST, and the lengths AC and ST. Because the interior angles of a triangle add up to 180°, you can find m∠A fairly easily:

$$m\angle A + m\angle B + m\angle C = 180°$$

$$m\angle A + 90° + 37° = 180°$$

$$m\angle A = 53°$$

Because corresponding angles are congruent, you know that m∠R = 53°, m∠S = 90° and m∠R = 37°. Using your proportionalities, you have

$$\frac{AB}{RS} = \frac{BC}{ST} = \frac{AC}{RT}$$

You can substitute the known values into the proportionality to find the value of ST:

$$\frac{3}{6} = \frac{4}{ST}$$

$$3(ST) = 24$$

$$ST = 8$$

To determine AC, you will use the other half of the proportionality:

$$\frac{3}{6} = \frac{AC}{10}$$

$$6(AC) = 30$$

$$AC = 5$$

You'll be using the idea that corresponding sides of similar triangles are proportional, so you might as well familiarize yourself with an abbreviation for that phrase now—it's CPSSAT.

The Big Three

In order to prove that two triangles are similar, you would need to verify that all three corresponding angles are congruent and that the required proportionality relationships hold between all corresponding sides. When you were working with congruent triangles you had some postulates and theorems to help you prove congruence. I'll give you some postulates and theorems to help you with similarity problems. Unfortunately, some of your similarity theorems have the same initials as the congruent triangle postulates. It's important to pay attention to whether you are trying to show that two triangles are similar or congruent. I'll throw the word "similarity" into any postulates or theorems just so you are clear on which one I'm using.

The AAA Similarity Postulate

Let me introduce you to your first shortcut involving the similarity of two triangles. It's a postulate, so it's something you can't prove. You will just have to believe in it and use it to your heart's content.

> **Postulate 13.1:** AAA Similarity Postulate. If the three angles of one triangle are congruent to the three angles of a second triangle, then the two triangles are similar.

This postulate lets you prove similarity without messing with the proportionalities. You only have to check the angle relationships. But you can even do better than that! If two angles of one triangle are congruent to two angles of another triangle, then the third angles must also be congruent. So if you want to show that two triangles are similar, all you have to do is show that two angles of one triangle are congruent to two angles of the other triangle.

Theorem 13.1: AA Similarity Theorem. If two angles of one triangle are congruent to two angles of a second triangle, then the two triangles are similar.

This theorem is easier to apply than the AAA Similarity Postulate (because you only have to check two angles instead of three). There's not much to the proof of Theorem 13.1. It relies mainly on fact that the measures of the interior angles of a triangle add up to 180°. Let's use it to prove the similarity of some triangles.

Example 7: If $\overline{AB} \parallel \overline{DE}$ as shown in Figure 13.4, write a two-column proof that shows △ABC ~ △EDC.

Figure 13.4

The segments $\overline{AB}$ *and* $\overline{DE}$ *are parallel.*

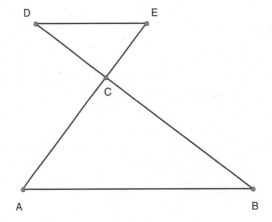

Solution: In order to write this proof, you need a game plan. The figures are getting a bit more complicated, and you have to use more and more of your previous results in order to write out proofs. Because $\overline{AB} \parallel \overline{DE}$, you can look at $\overline{AB}$ and $\overline{DE}$ as two parallel lines cut by a transversal $\overline{AE}$. In this case, ∠BAE and ∠DEA are alternate interior angles, so they are congruent. Because ∠ACB and ∠DCE are vertical angles, they are congruent. So by the AA Similarity Theorem, you see that △ABC ~ △EDC.

Statements	Reasons
1. $\overline{AB} \parallel \overline{DE}$	Given
2. $\overline{AB}$ and $\overline{DE}$ as two parallel lines cut by a transversal $\overline{AE}$	Definition of transversal
3. ∠BAE and ∠DEA are alternate interior angles	Definition of alternate interior angles

Statements	Reasons
4. ∠BAE ≅ ∠DEA	Theorem 10.8
5. ∠ACB and ∠DCE are vertical angles	Definition of vertical angles
6. ∠ACB ≅ ∠DCE	Theorem 8.1
7. ΔABC ~ ΔEDC	AA Similarity Theorem

One technique for estimating the height of an object (like a tree, or a pyramid) uses the ideas of similar triangles. This technique assumes that you know your own height and can measure the lengths of shadows. In order for this technique to work, both you and the object you are trying to measure must cast a shadow.

Suppose that the sun is shining, and you want to determine the height of a nearby tree. In order for this technique to work, the sun can't be shining directly overhead—otherwise neither you nor the tree will cast a measurable shadow. Figure 13.5 shows the role that similar triangles play in this technique. Suppose you are 6 feet tall, and you cast a shadow of length 8 feet. You don't know how tall the tree is, but its shadow is 36 feet long. If you assume that both you and the tree have good posture and stand perpendicular to the ground, both you and the tree form two triangles. Because the sun is very far away, you can assume that ∠A and ∠D are congruent. Using your AA Similarity Theorem, you can show that ΔABC ~ ΔEDC.

Using the idea that CSSTAP, we see that

$$\frac{AC}{BC} = \frac{DF}{EF} \text{ or } \frac{6}{8} = \frac{h}{36}.$$

Cross-multiply and we see that

$$h = \frac{6 \times 36}{8} = 27.$$

So the tree is roughly 27 feet tall. Thales used this method to estimate the height of the pyramids, and he was accurate enough to have amazed his friends and impressed some pharaohs.

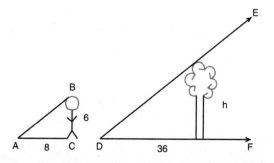

Figure 13.5

The stick figure and the tree cast a shadow. Two similar triangles are formed, and because the sides are proportional, we can determine the height of the tree.

The SAS and SSS Similarity Theorems

There are other theorems that can help show that two triangles are similar. I will just state two other theorems that can be useful. I won't take the time to prove these theorems, because there's so much to discuss and I'm running out of space.

> **Theorem 13.2:** The SAS Similarity Theorem. If an angle of one triangle is congruent to an angle of a second triangle and the including sides of that angle are proportional, then the triangles are similar.

> **Theorem 13.3:** The SSS Similarity Theorem. If three sides of one triangle are proportional to the three corresponding sides of another triangle, then the triangles are similar.

Example 8: Given the two triangles shown in Figure 13.6, find RS.

Figure 13.6

$\triangle ABC$ *and* $\triangle RST$.

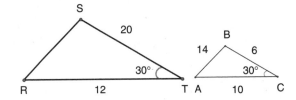

Solution: We see that $\angle C \cong \angle T$ and that the pairs of including sides are proportional:

$$\frac{6}{10} = \frac{12}{20}$$

Don't be fooled into thinking that the two sides are not proportional because

$$\frac{6}{10} \neq \frac{20}{12}.$$

You have to try various combinations to determine how the sides correspond. In this case, $\overline{AC}$ corresponds to $\overline{ST}$, $\overline{BC}$ corresponds to $\overline{RT}$, and $\overline{AB}$ corresponds to $\overline{RS}$. So $\triangle ABC \sim \triangle RST$ by the SAS Similarity Theorem, and you can use the fact that CSSTAP to determine RS:

$$\frac{RS}{ST} = \frac{AB}{AC}$$

$$\frac{RS}{20} = \frac{14}{10}$$

$$RS = 28$$

Put Me in, Coach!

Here's your chance to shine. Remember that I am with you in spirit and have provided the answers to these questions in Appendix A.

1. Use the Means-Extremes Property to solve for x: $\dfrac{x-2}{3} = \dfrac{x+1}{4}$.

2. Use the Means-Extremes Property to solve for x: $\dfrac{6}{x} = \dfrac{x}{24}$.

3. Suppose that the measures of two supplementary angles are in the ratio of 2 to 7. Find the measure of each angle.

4. If $\angle A \cong \angle D$ as shown in Figure 13.7, write a two-column proof to show that $\dfrac{BC}{AB} = \dfrac{CE}{DE}$.

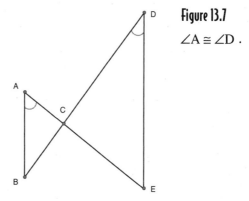

Figure 13.7

$\angle A \cong \angle D$.

5. The distance across an alligator-infested lake is to be measured safely by using similar triangles, as shown in Figure 13.8. If XY = 160 feet, YW = 40 feet, TY = 120 feet, and WZ = 50 feet, find XT.

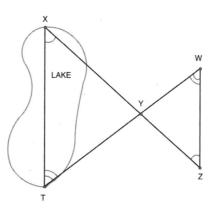

Figure 13.8

Measuring the length of a lake using similar triangles.

The Least You Need to Know

◆ The Means-Extremes property of proportions is often used when working with proportions.

◆ The AAA Similarity Postulate is the starting point for proving that two triangles are similar.

◆ Of the three theorems used to prove that two triangles are similar (AA, SAS and SSS Similarity Theorems), AA is the most useful.

◆ When you know that two triangles are similar, use the fact that CSSTAP to determine the lengths of sides and angles of the triangles.

14

Opening Doors with Similar Triangles

In This Chapter

◆ The first proof of the Pythagorean Theorem

◆ Dividing segments proportionally

◆ Using similar triangles to prove lines are parallel

◆ Special triangles: 60-60-60, 30-60-90, and 45-45-90

You just spent a whole chapter talking about similar triangles, but you didn't actually do much with them. Oh sure, I showed you how to use theorems to prove that two triangles were similar, and you were able to measure the length of a lake using similar triangles, but there have to be more applications than that.

One of the most useful things about similar triangles is CSSTAP (corresponding sides of similar triangles are proportional). After you show that two triangles are similar, you can develop proportionalities that will help you establish relationships between the sides of the similar triangles.

In addition to using the properties of similar triangles to measure lake dimensions, you can prove the Pythagorean Theorem and show that two lines are parallel. And that's just the beginning.

The Pythagorean Theorem

Now that you know how to show that two triangles are similar, you can use CSSTAP to find relationships between the sides of similar triangles. You can even create the theorems necessary to prove one of the most famous theorems in geometry: the Pythagorean Theorem. But before you can tackle the Pythagorean Theorem, you'll need a theorem about altitudes. I'll walk through an explanation of why the theorem is true, but I will not write out a formal proof.

> **Theorem 14.1**: The altitude drawn to the hypotenuse of a right triangle separates the right triangle into two right triangles that are similar to each other and to the original right triangle.

Figure 14.1 will help clarify what is going on. The altitude drawn to the hypotenuse has to originate at the vertex of the right angle (C) and is perpendicular to the hypotenuse ($\overline{AB}$). Let's call the point where the altitude and the hypotenuse intersect D. You have three triangles to relate: the original triangle $\triangle ABC$ and two new, smaller triangles $\triangle ACD$ and $\triangle CBD$. Remember, you have to be careful about the order of the vertices. You have to match up corresponding vertices when representing the similarity of two triangles. You can use the AA Similarity Theorem and the transitive property of similarity (remember, similarity is an equivalence relation) to show that all three triangles are similar.

Figure 14.1

$\triangle ABC$ *with altitude* $\overline{CD}$
from C *to the hypotenuse*
$\overline{AB}$.

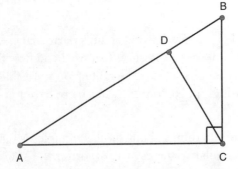

First, I'll show that $\triangle ABC$ and $\triangle CBD$ are similar. Both triangles are right triangles, so there's one pair of congruent angles. As for the second angle, notice that $\angle B$ is involved in both triangles. Because $\angle B \cong \angle B$, you've got a second pair of congruent angles, and by the AA Similarity Theorem, $\triangle CBD \cong \triangle ABC$. You need to match up your angles and your vertices: $\angle D$ is the right angle in $\triangle CBD$, which corresponds to $\angle C$ in $\triangle ABC$, $\angle B$ corresponds to itself, and $\angle BCD$ corresponds to $\angle A$.

You'll use a similar argument to show that $\triangle ABC$ and $\triangle ACD$ are similar. $\angle D$ is the right angle in $\triangle ACD$, so it corresponds to $\angle C$ in $\triangle ABC$. $\angle A$ corresponds to itself, which leaves $\angle ACD$ and $\angle B$ to correspond.

Because $\triangle ACD \sim \triangle ABC$ and $\triangle ABC \sim \triangle CBD$, the transitive property of $\sim$ shows that $\triangle ACD \sim \triangle CBD$. Because CSSTAP, you know that

$$\frac{AD}{CD} = \frac{CD}{DB}.$$

This should look a bit familiar. From this proportionality, you see that CD is the geometric mean of the lengths of the segments of the hypotenuse! The Pythagorean Theorem is just a couple of algebraic steps away.

Now that you know these three triangles are similar, you can break them apart to get a better handle on the proportionalities involved. You might want to refer to Figure 14.2.

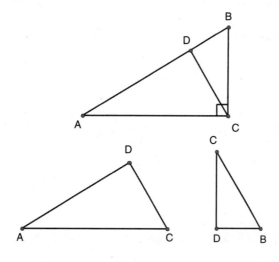

Figure 14.2

$\triangle ABC$ can be split into two triangles: $\triangle ACD$ and $\triangle CBD$. All three triangles are similar.

Because $\triangle ACD \sim \triangle ABC$, you know that

$$\frac{AB}{AC} = \frac{AC}{AD}.$$

Because $\triangle ABC \sim \triangle CBD$, you know that

$$\frac{AB}{BC} = \frac{BC}{BD}.$$

If you cross-multiply, add the two equations together, and simplify, you will derive the Pythagorean Theorem! I'll write out the details in a two-column proof.

The Pythagorean Theorem: The square of the length of the hypotenuse of a right triangle is equal to the sum of the squares of the lengths of the legs.

Translating the Pythagorean Theorem into the drawing in Figure 14.2, the goal is to prove that $(AB)^2 = (BC)^2 + (AC)^2$.

	Statements	Reasons
1.	$\triangle ACD \sim \triangle ABC$ and $\triangle ABC \sim \triangle CBD$	Previous discussion
2.	$\dfrac{AB}{AC} = \dfrac{AC}{AD}$ and $\dfrac{AB}{BC} = \dfrac{BC}{BD}$	CSSTAP
3.	$AD + BD = AB$	Segment Addition Postulate
4.	$(AB)(AD) = (AC)^2$, $(AB)(BD) = (BC)^2$	Means-Extremes property of proportionality
5.	$(AB)(AD) + (AB)(BD) = (BC)^2 + (AC)^2$	Addition property of equality
6.	$(AB)[(AD) + (BD)] = (BC)^2 + (AC)^2$	Algebra (factoring)
7.	$(AB)[(AB)] = (BC)^2 + (AC)^2$	Substitution (step 3 and 6)
8.	$(AB)^2 = (BC)^2 + (AC)^2$	Algebra

That's the most algebraic proof of the Pythagorean Theorem! The next time you prove the Pythagorean Theorem you will use areas of triangles.

Parallel Segments and Segment Proportions

Suppose you have a segment $\overline{AB}$. You've talked about breaking up a segment into two equal pieces (using the midpoint). You can also break up a segment into thirds, quarters, or whatever fraction you want.

Suppose you have two segments, $\overline{AB}$ and $\overline{RS}$, as shown in Figure 14.3. These segments can have the same length, or they can have different lengths. You can break each segment up into a variety of pieces. One specific way you can divide up your segments is proportionally. When two segments, $\overline{AB}$ and $\overline{RS}$, are divided proportionally, it means that you have found two points, C on $\overline{AB}$ and T on $\overline{RS}$, so that

$$\frac{AC}{RT} = \frac{CB}{TS}.$$

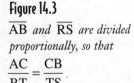

Figure 14.3

$\overline{AB}$ and $\overline{RS}$ are divided proportionally, so that

$$\frac{AC}{RT} = \frac{CB}{TS}$$

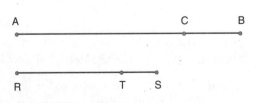

You are now ready to prove the following theorem:

Theorem 14.2: If a line is parallel to one side of a triangle and intersects the other two sides, then it divides these sides proportionally.

Example 1: Write a formal proof of Theorem 14.2.

Solution: This theorem is illustrated in Figure 14.4.

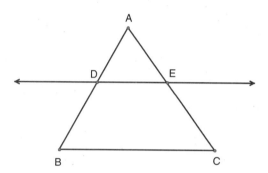

Figure 14.4

$\triangle ABC$ *has* $\overleftrightarrow{DE} \| \overline{BC}$, *with* $\overleftrightarrow{DE}$ *intersecting* $\overline{AB}$ *at D and* $\overline{AC}$ *at E.*

Given: In Figure 14.4, $\triangle ABC$ has $\overleftrightarrow{DE} \| \overline{BC}$, with $\overleftrightarrow{DE}$ intersecting $\overline{AB}$ at D and $\overline{AC}$ at E.

Prove: $\dfrac{AD}{DB} = \dfrac{AE}{EC}$.

Proof: In order to show that D and E divide the segments $\overline{AB}$ and $\overline{AC}$ proportionally, you will need to show that $\triangle ADE \sim \triangle ABC$ and then use CSSTAP. To show that $\triangle ADE \sim \triangle ABC$, you will use the AA Similarity Theorem. To determine the angle congruencies, you will use our postulate about corresponding angles and parallel lines.

Statements	Reasons
1. $\triangle ABC$ has $\overleftrightarrow{DE} \| \overline{BC}$, with $\overleftrightarrow{DE}$ intersecting $\overline{AB}$ at D and $\overline{AC}$ at E	Given
2. $\overleftrightarrow{DE} \| \overline{BC}$ cut by a transversal $\overline{AB}$	Definition of transversal
3. $\angle ADE$ and $\angle ABC$ are corresponding angles	Definition of corresponding angles
4. $\angle ADE \cong \angle ABC$	Postulate 10.1
5. $\angle DAE \cong \angle BAC$	Reflexive property of $\cong$
6. $\triangle ADE \sim \triangle ABC$	AA Similarity Theorem
7. $\dfrac{AB}{AD} = \dfrac{AC}{AE}$	CSSTAP

Statements	Reasons
8. $\dfrac{AB - AD}{AD} = \dfrac{AC - AE}{AE}$	Property 3 of proportionalities
9. $\dfrac{BD}{AD} = \dfrac{EC}{AE}$	Segment Addition Postulate
10. $\dfrac{AD}{BD} = \dfrac{AE}{EC}$	Property 2 of proportionalities

You can also use similar triangles to show that two lines are parallel. For example, suppose that $\triangle ADE \sim \triangle ABC$ in Figure 14.5. You can prove that $\overline{DE} \,||\, \overline{BC}$.

Example 2: If $\triangle ADE \sim \triangle ABC$ as shown in Figure 14.5, prove that $\overline{DE} \,||\, \overline{BC}$.

Figure 14.5

$\triangle ADE \sim \triangle ABC$.

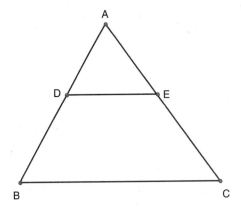

Solution: Your game plan is quite simple. Because $\triangle ADE \sim \triangle ABC$, you know that $\angle ADE \cong \angle ABC$. Because $\angle ADE$ and $\angle ABC$ are congruent corresponding angles, you know that $\overleftrightarrow{DE} \,||\, \overleftrightarrow{BC}$ by Theorem 10.7.

Statements	Reasons
1. $\triangle ADE \sim \triangle ABC$	Given
2. $\angle ADE \cong \angle ABC$	Definition of $\sim$
3. $\overleftrightarrow{DE}$ and $\overleftrightarrow{BC}$ are two lines cut by a transversal $\overleftrightarrow{AB}$	Definition of transversal
4. $\angle ADE$ and $\angle ABC$ are corresponding angles	Definition of corresponding angles
5. $\overline{DE} \,||\, \overline{BC}$	Theorem 10.7

Three Famous Triangles

Now that you have developed the skills needed to prove either that two triangles are congruent or two triangles are similar, you can explore the properties of three of the most famous types of triangles: the 45-45-90 triangle, the 30-60-90 triangle, and the 60-60-60 triangle. Start with the equiangular triangle first.

60-60-60 Triangle

A 60-60-60 triangle is an equiangular triangle. If you look at the equiangular triangle in Figure 14.6, some observations are immediate. If you flip the triangle so that vertex A corresponds with vertex B, and vertex B corresponds to vertex A, then by the AA Similarity Theorem, $\triangle ABC \sim \triangle BAC$. So what? Well, because CSSTAP, you see that

$$\frac{BC}{AB} = \frac{AC}{AB}$$, so $BC = AC$.

You can flip the triangle again, so that vertex A corresponds to vertex C and vertex C corresponds to vertex A. Then using CSSTAP you have the proportionality

$$\frac{AB}{AC} = \frac{BC}{AC}$$, or that $AB = BC$.

No matter how you slice it, all three sides of an equiangular triangle are congruent. So an equiangular triangle is an equilateral triangle, and an equilateral triangle is an equiangular triangle. I will state this as a theorem.

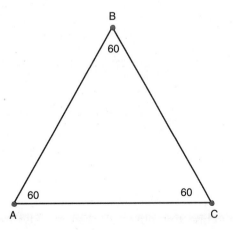

Figure 14.6

$\triangle ABC$ *is an equiangular triangle.*

Theorem 14.3: Every equiangular triangle is an equilateral triangle, and every equilateral triangle is an equiangular triangle.

Now that you know that a 60-60-60 triangle is also an equilateral triangle, it's time to bisect this triangle and see what it's made of. Figure 14.7 shows an equiangular triangle $\triangle ABC$, with an altitude dropped from vertex B to side $\overline{AC}$. Call the point of intersection of the altitude and $\overline{AC}$ point D. Again, using the AA Similarity Theorem, you can show that $\triangle ABD \sim \triangle CBD$. You can use the properties of similar triangles to show that

Eureka! _____

Every equiangular triangle is an equilateral triangle, and every equilateral triangle is an equiangular triangle!

$$\frac{AD}{AB} = \frac{DC}{BC}, \text{ or that } AD = DC.$$

Because AD + DC = AC (by the Segment Addition Postulate), you see that D is the midpoint of $\overline{AC}$ and that

$$AD = \tfrac{1}{2}AC.$$

Figure 14.7

$\triangle ABC$ _with altitude_ $\overrightarrow{BD}$ _from_ B _to_ $\overline{AC}$.

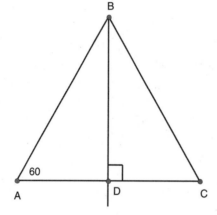

The interior angles of a triangle add up to 180°. If m∠BAD = 60° and m∠BDA = 90°, then m∠DBA = 30°. Because $\triangle ABD \sim \triangle CBD$, you also know that m∠DBC = 30°, and that $\overrightarrow{BD}$ (the altitude from B to $\overline{AC}$) bisects ∠ABC. You have also just created the second special triangle.

30-60-90 Triangle

The second special triangle is the 30-60-90 triangle. It can be created by taking an equilateral triangle and bisecting one angle. Figure 14.8 shows two 30-60-90 triangles formed by bisecting an equilateral triangle. These two triangles formed are congruent, and are worth further investigation.

Each of these triangles has two legs: one formed from one-half of the side of the equilateral triangle (either $\overline{AD}$ or $\overline{DC}$, depending on the triangle) and the other formed by the altitude from vertex B to $\overline{AC}$ ($\overline{BD}$). You can use the Pythagorean Theorem to discover relationships between the lengths of the sides. Let's focus on $\triangle ABD$: If you let x represent AD, then AB = 2x, as shown in Figure 14.8. This is because all sides of an equilateral triangle are congruent, and when you divided $\triangle ABC$ into two triangles, one of the sides was cut in half. Using the Pythagorean Theorem, you can find the length of the second leg:

$$(AD)^2 + (BD)^2 = (AB)^2$$

$$(x)^2 + (BD)^2 = (2x)^2$$

$$(BD)^2 = 4x^2 - x^2 = 3x^2$$

$$BD = x\sqrt{3}$$

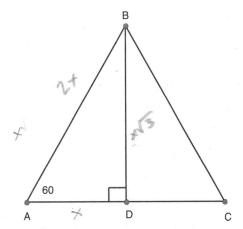

Figure 14.8

$\triangle ABD$ *is a 30-60-90 triangle with* AD = x.

Because $\sqrt{3} > 1$, you can see that the length of the leg opposite the 60° angle is bigger than the length of the leg opposite the 30° angle. I will state these results as a theorem:

Theorem 14.4: In a 30-60-90 triangle, the hypotenuse has a length equal to twice the length of the leg opposite the shorter angle, and the length of the leg opposite the 60° angle is the product of $\sqrt{3}$ and the length of the shorter leg.

Eureka!

For a 30-60-90 triangle, if the side opposite the smallest angle has length **x** (it is the smallest side), then the side opposite the middle angle has length $x\sqrt{3}$ (it is the middle side) and the hypotenuse has length **2x** (and is the longest side).

You can use these results to find the missing pieces.

Example 3: Find the lengths of the missing sides of each triangle in Figure 14.9.

Figure 14.9

$\triangle$ABC *and* $\triangle$RST.

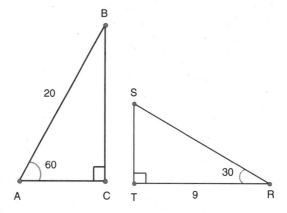

Solution: Let's look at $\triangle$ABC first. Because the length of the hypotenuse is given to be 20, you know that the shorter leg has one-half this length. The shorter leg is the leg opposite the 30° angle, which is $\overline{AC}$. So AC = 10. The other leg (the one opposite the 60° angle) has length equal to the product of the length of the shorter leg and $\sqrt{3}$, so BC $= 10\sqrt{3}$.

For $\triangle$RST, things are a little trickier. You are given the length of the leg opposite the 60° angle, and you need to find the length of the shorter leg and the hypotenuse. Because RT $=$ ST$\sqrt{3}$, and RT = 9, you have that $9 =$ ST$\sqrt{3}$, or ST $= \dfrac{9}{\sqrt{3}} = 3\sqrt{3}$. The hypotenuse has length equal to twice the length of the shortest leg, so RS $= 2 \times 3\sqrt{3} = 6\sqrt{3}$.

45-45-90 Triangle

A 45-45-90 triangle has two acute angles with equal measure and one right angle. You can use the Pythagorean Theorem to find the relationships between the lengths of the legs and the length of the hypotenuse.

Figure 14.10 shows a typical 45-45-90 triangle. You can easily prove that the two legs are congruent by showing $\triangle$ABC $\cong$ $\triangle$BAC (using the SAS Postulate).

You can combine this result with the Pythagorean Theorem to determine the relationship between the length of one of the legs and the length of the hypotenuse. Because $(AB)^2 = (BC)^2 + (AC)^2$, and AC = BC, you have $(AB)^2 = 2(BC)^2$. Using the Square Root Property of Equality, you see that AB $=$ BC$\sqrt{2}$.

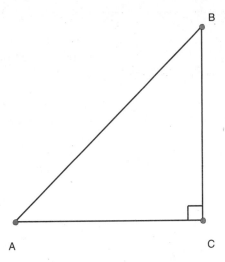

Figure 14.10

A typical 45-45-90 triangle.

You can use these relationships to find the missing pieces of the triangle.

Eureka!

The length of the hypotenuse of a 45-45-90 triangle can be found by multiplying the length of the leg by $\sqrt{2}$.

Example 4: Find the lengths of the missing sides in the triangle shown in Figure 14.11.

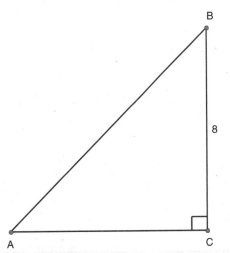

Figure 14.11

$\triangle ABC$ *is a 45-45-90 triangle.*

Solution: Because the length of one of the legs is 8, the length of the hypotenuse is $8\sqrt{2}$ and the length of the other leg is 8.

Put Me in, Coach!

Here's your chance to shine. Remember that I am with you in spirit and have provided the answers to these questions in Appendix A.

1. Prove that if a line is parallel to one side of a triangle and passes through the midpoint of a second side, then it will pass through the midpoint of the third side.

2. Find the lengths of the missing sides of each triangle in Figure 14.12.

Figure 14.12

△ABC *and* △RST.

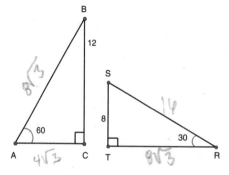

3. Find the lengths of the missing sides in the triangles shown in Figure 14.13.

Figure 14.13

△ABC *is a 45-45-90 triangle.*

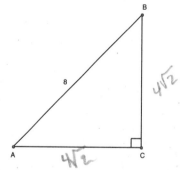

The Least You Need to Know

◆ You saw the first three proofs of the Pythagorean Theorem.

◆ Use similar triangles to prove that two lines are parallel.

◆ Use similar triangles to compute the lengths of sides using CSSTAP.

◆ Calculate the side lengths of a 30-60-90 triangle and a 45-45-90 triangle.

Chapter 15

Putting Quadrilaterals in the Forefront

In This Chapter

- ◆ Properties of quadrilaterals
- ◆ Learn about trapezoids
- ◆ Study the different kinds of parallelograms
- ◆ Calculate the area of quadrilaterals

If you look around your house, the most common polygon you will see is a quadrilateral. Your walls, windows, doors, floors, and ceilings are probably quadrilaterals. There must be something special about them if these four-sided polygons form the basis for most of our building designs.

There are many different quadrilaterals. Squares and rectangles are probably the most popular, but many people appreciate kites, parallelograms, rhombuses, and trapezoids. Each of these quadrilaterals has its own peculiar combination of properties that warrant giving it its own special name.

Before you started reading this book, I'm sure you could pick out a square in a quadrilateral line-up. But you've probably never written a theorem (the mathematical equivalent of a poem) about a square. That's all about to change.

Properties of All Quadrilaterals

Quadrilaterals: The shape of books, envelopes, picture frames, and more. Let's think about the properties that all quadrilaterals possess.

First of all, all quadrilaterals have four sides. There is no such thing as a three-sided quadrilateral. Don't confuse a three-sided quadrilateral with a three-sided polygon. A three-sided polygon is called a triangle, and you just spent several chapters exploring its properties. A three-sided quadrilateral does not exist, even in the twilight zone.

Every quadrilateral has two diagonals, and the sum of the measures of the interior angles is 360°. You calculated this in Chapter 6.

Now it's time to explore the properties of the various types of quadrilaterals. For example, you will explore what makes a rectangle a rectangle, and not a kite. I'll start with the most general quadrilaterals and end with the most restrictive shapes.

Properties of Trapezoids

A *trapezoid* is a quadrilateral with exactly two parallel sides. Figure 15.1 shows trapezoid ABCD. Remember the naming conventions for polygons. You must list the vertices in consecutive order. In trapezoid ABCD, $\overline{BC} \parallel \overline{AD}$. The parallel sides $\overline{BC}$ and $\overline{AD}$ are called the *bases*, and the nonparallel sides $\overline{AB}$ and $\overline{CD}$ are *legs*. *Base angles* are a pair of angles that share a common base. In Figure 15.1, $\angle A$ and $\angle D$ form one set of base angles.

Figure 15.1

The trapezoid ABCD.

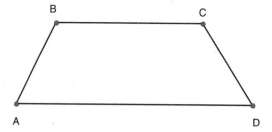

When the midpoints of the two legs of a trapezoid are joined together, the resulting segment is called the *median* of the trapezoid. In Figure 15.2, R and S are the midpoints of $\overline{AB}$ and $\overline{CD}$, and $\overline{RS}$ is the median of trapezoid ABCD. The median of a trapezoid is parallel to each base. Strangely enough, the length of the median of a trapezoid equals one-half the sum of the lengths of the two bases. Accept these statements as theorems (without proof), and use them when needed.

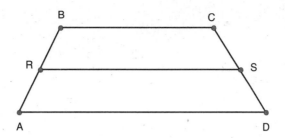

Figure 15.2

R *and* S *are the midpoints of* $\overline{AB}$ *and* $\overline{CD}$, *and* $\overline{RS}$ *is the median of trapezoid* ABCD.

Theorem 15.1: The median of a trapezoid is parallel to each base.

Theorem 15.2: The length of the median of a trapezoid equals one-half the sum of the lengths of the two bases.

Example 1: In trapezoid ABCD, $\overline{BC} \parallel \overline{AD}$, R is the midpoint of $\overline{AB}$ and S is the midpoint of $\overline{CD}$, as shown in Figure 15.3. Find AD, BC, and RS if BC = 2x, RX = 4x − 25 and AD = 3x − 5.

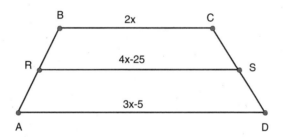

Figure 15.3

Trapezoid ABCD, $\overline{BC} \parallel \overline{AD}$ $\overline{AB}$ *has midpoint* R *and* $\overline{CD}$ *has midpoint* S.

Solution: Because RS = ½(AD + BC), you can substitute the values for each segment length:

4x − 25 = ½(3x − 5 + 2x)

Rearranging and simplifying gives:

$$4x - 25 = \frac{5}{2}x - \frac{5}{2}$$

$$4x - \frac{5}{2}x = 25 - \frac{5}{2}$$

$$\frac{3}{2}x = \frac{45}{2}$$

x = 15

So, x = 15, BC = 30, RS = 35, and AD = 40.

An *altitude* of a trapezoid is a perpendicular line segment from a vertex of one base to the other base (or to an extension of that base). In Figure 15.4, $\overline{BT}$ is an altitude of trapezoid ABCD.

Figure 15.4

The trapezoid ABCD, *with altitude* $\overline{BT}$.

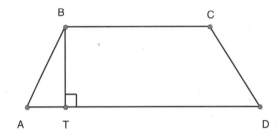

Solid Facts

A **trapezoid** is a quadrilateral with exactly two parallel sides.

The **bases** of a trapezoid are the parallel sides.

The **legs** of a trapezoid are the nonparallel sides.

The **median** of a trapezoid is the line segment joining the midpoints of the two legs.

An **altitude** of a trapezoid is a perpendicular line segment from a vertex of one base to the other base (or to an extension of that base).

Base angles of a trapezoid are a pair of angles that share a common base.

Built into the trapezoid are two parallel lines (the bases $\overline{BC}$ and $\overline{AD}$) cut by a transversal (one of the legs, either $\overline{AB}$ or $\overline{CD}$). You know that the two interior angles on the same side of the transversal are supplementary angles (Theorem 10.5), so $\angle A$ and $\angle B$ are supplementary angles, as are $\angle C$ and $\angle D$.

Let's All Fly a Kite!

Although there are many designs for kites, my kites all look like the kite in Figure 15.5. These kites are constructed by attaching two sticks of different lengths together so that the sticks are perpendicular and one of the sticks bisects the other. In this construction, there are two pairs of congruent adjacent sides. And that is what makes a kite a kite. A *kite* is defined as a quadrilateral with two distinct pairs of congruent adjacent sides.

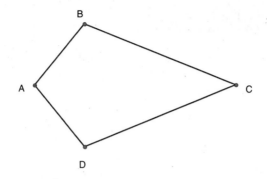

Figure 15.5
The kite ABCD.

In a kite, one pair of opposite angles are congruent. Also, in a kite, the diagonals are perpendicular and one of them bisects the other. I'll walk you through the proofs of these statements and give you the opportunity to write the proof itself.

> **Theorem 15.3:** In a kite, one pair of opposite angles are congruent.
>
> **Theorem 15.4:** The diagonals of a kite are perpendicular, and the diagonal opposite the congruent angles bisects the other diagonal.

Solid Facts

A **kite** is a quadrilateral with two distinct pairs of congruent adjacent sides.

You can prove Theorem 15.3 by using the SSS Postulate. The kite ABCD has $\overline{AB} \cong \overline{AD}$ and $\overline{BC} \cong \overline{CD}$, and the reflexive property of $\cong$ enables you to write $\overline{AC} \cong \overline{AC}$. Then by CPOCTAC, you have $\angle B \cong \angle D$. As an added bonus, you also have $\angle BAC \cong \angle DAC$. This will come in handy when proving Theorem 15.4.

To prove Theorem 15.4, suppose that the two diagonals intersect at M, as shown in Figure 15.6. Consider the two small triangles formed: $\triangle AMD$ and $\triangle AMB$. Because $\overline{AB} \cong \overline{AD}$, $\overline{AM} \cong \overline{AM}$, and $\angle BAM \cong \angle DAM$, you can use the SAS Postulate to show $\triangle AMD \cong \triangle AMB$. Thus $\overline{BM} \cong \overline{MD}$ (and you've got bisection of one of the diagonals) and $\angle BMA \cong \angle DMA$ (and you have right angles).

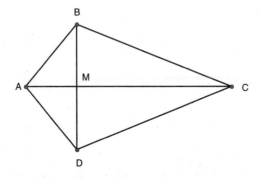

Figure 15.6
Kite ABCD.

Properties of Parallelograms

A *parallelogram* is a quadrilateral that has both pairs of opposite sides parallel. Parallelograms have many properties that are easy to prove using the properties of parallel lines. You will occasionally use a diagonal to divide a parallelogram into triangles. If you do this carefully, your triangles will be congruent, so you can use CPOCTAC.

> **Theorem 15.5**: A diagonal of a parallelogram separates it into two congruent triangles.

Solid Facts

A **parallelogram** is a quadrilateral that has both pairs of opposite sides parallel.

Example 2: Write a formal proof of Theorem 15.5.

Solution: Begin by going down the list of what you need to bring to a formal proof. We already have the statement of the theorem. Figure 15.7 shows parallelogram ABCD with diagonal $\overline{AC}$.

Figure 15.7

Parallelogram ABCD *with diagonal* $\overline{AC}$.

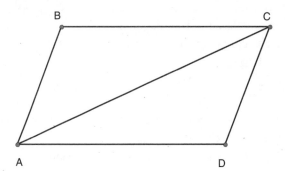

Given: Parallelogram ABCD with diagonal $\overline{AC}$.

Prove: $\triangle ABC \cong \triangle CDA$.

Proof: Your game plan is to make use of the properties of parallel lines cut by a transversal to relate two of the angles of $\triangle ABC$ with two corresponding angles in $\triangle CDA$. Because $\overline{AC} \cong \overline{AC}$, you can use the ASA Postulate to show $\triangle ABC \cong \triangle CDA$.

Statements	Reasons
1. Parallelogram ABCD has diagonal $\overline{AC}$	Given
2. $\overline{BC} \parallel \overline{AD}$ cut by transversal $\overline{AC}$	Definition of transversal
3. ∠BAC and ∠DCA are alternate interior angles	Definition of alternate interior angles

Statements	Reasons
4. ∠BAC ≅ ∠DCA	Theorem 10.2
5. $\overline{BC} \parallel \overline{AD}$ cut by transversal $\overline{AC}$	Definition of transversal
6. ∠ACB and ∠DAC are alternate interior angles	Definition of alternate interior angles
7. ∠ACB ≅ ∠DAC	Theorem 10.2
8. $\overline{AC} \cong \overline{AC}$	Reflexive property of ≅
9. ∆ABC ≅ ∆CDA	ASA Postulate

This theorem will come in handy when establishing theorems about parallelograms. A common technique involves using a diagonal to divide a parallelogram into two triangles and then applying CPOCTAC. The next two theorems use this technique. I'll prove the first one and let you prove the second.

Theorem 15.6: Opposite sides of a parallelogram are congruent.

Theorem 15.7: Opposite angles of a parallelogram are congruent.

Example 3: Write a two-column proof of Theorem 15.6.

Solution: You can draw from the information shown in Figure 15.7. We'll be dealing with the parallelogram ABCD and its diagonal $\overline{AC}$. You will want to prove $\overline{BC} \cong \overline{AD}$.

Statements	Reasons
1. Parallelogram ABCD has diagonal $\overline{AC}$	Given
2. ∆ABC ≅ ∆CDA	Theorem 15.5
3. $\overline{BC} \cong \overline{AD}$	CPOCTAC

The last property of a parallelogram that I will mention involves the intersection of the diagonals. It turns out that the diagonals of a parallelogram bisect each other. The proof of this is fairly straightforward, so I'll walk you through the game plan and let you provide the details.

Theorem 15.8: The diagonals of a parallelogram bisect each other.

Take a look at parallelogram ABCD in Figure 15.8. It has diagonals $\overline{AC}$ and $\overline{BD}$ which intersect at M. We want to show $\overline{AM} \cong \overline{MC}$. The easiest way to do this is to find two triangles that are congruent and use CPOCTAC. The two triangles that we'll try to prove congruent are ∆AMD and ∆CMB. Because opposite sides of a parallelogram are congruent, $\overline{BC} \cong \overline{AD}$. Because vertical angles are congruent, $\angle AMD \cong \angle CMB$. Finally, we have $\overline{BC} \parallel \overline{AD}$ cut by a transversal $\overline{AC}$, and because $\angle BCA$ and $\angle CAD$ are alternate interior angles, they are congruent. Using the AAS Theorem, we can conclude that $\triangle AMD \cong \triangle CMB$. Finish it up by using CPOCTAC.

Figure 15.8

Parallelogram ABCD *has diagonals* $\overline{AC}$ *and* $\overline{BD}$ *which intersect at* M.

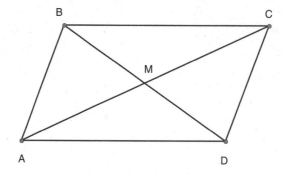

The Most Popular Parallelograms

Rectangles, rhombuses, and squares are three specific kinds of parallelograms. They all have the properties of a parallelogram: Their opposite sides are parallel, their diagonals bisect each other and divide the parallelogram into two congruent triangles, and opposite sides and angles are congruent. But rectangles, rhombuses, and squares have additional characteristics that other parallelograms don't have.

Rectangles

A *rectangle* is a parallelogram that has a right angle. Actually, from this little bit of information, you know about all four angles of a rectangle. A rectangle is a parallelogram, so its opposite angles are congruent and its consecutive angles are supplementary. Recall that the supplement of a right angle is another right angle. So a rectangle actually has four right angles.

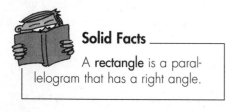

Solid Facts

A **rectangle** is a parallelogram that has a right angle.

Rectangles have some properties that generic parallelograms do not. One such property is that the diagonals of a rectangle are congruent. I will state that as a theorem and discuss a game plan for the proof. I will leave the details up to you.

Theorem 15.9: The diagonals of a rectangle are congruent.

To prove this theorem, take a look at the rectangle in Figure 15.9. Rectangle ABCD has diagonals $\overline{AC}$ and $\overline{BD}$. In order to prove that they are congruent, you will want to use CPOCTAC. But which two triangles do you show are congruent? I would recommend that you show $\triangle ADC \cong \triangle DAB$. Split these two triangles apart and make the connections. They are both right triangles, so that's one pair of congruent angles. Because opposite sides are congruent, you will be able to use our SAS Postulate to show $\triangle ADC \cong \triangle DAB$.

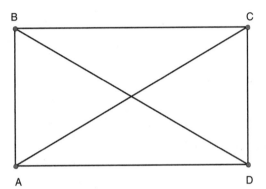

Figure 15.9

Rectangle ABCD *has diagonals* $\overline{AC}$ *and* $\overline{BD}$.

Rhombuses

A *rhombus* is a parallelogram with two congruent adjacent sides. Just as you saw with a rectangle, a rhombus inherits all of the desirable properties of a parallelogram. And rhombuses have special properties that generic parallelograms and rectangles do not have. I'll write these special properties as theorems that you can refer to later.

The first property of a rhombus is that all sides of a rhombus are congruent. That is not surprising, because you already know that the opposite sides of a rhombus are congruent (because it's a parallelogram). If opposite sides and adjacent sides are congruent, they are all congruent. Not all parallelograms and rectangles have this special property.

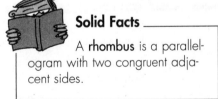

Solid Facts

A **rhombus** is a parallelogram with two congruent adjacent sides.

   Theorem 15.10: All sides of a rhombus are congruent.

The next property of a rhombus worth mentioning is that its diagonals are perpendicular. That one might not be so obvious, and is worth writing out a formal proof.

Theorem 15.11: The diagonals of a rhombus are perpendicular.

Example 4: Write a formal proof of Theorem 15.11.

Solution: As usual, I'll write the formal proof step-by-step. Figure 15.10 shows the rhombus ABCD with diagonals $\overline{AC}$ and $\overline{BD}$ that intersect at M.

Figure 15.10

Rhombus ABCD *with diagonals* $\overline{AC}$ *and* $\overline{BD}$ *that intersect at* M.

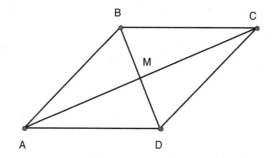

Given: Rhombus ABCD with diagonals $\overline{AC}$ and $\overline{BD}$.

Prove: $\overline{AC} \perp \overline{BD}$.

Proof: You need a game plan. Hang on to your hat because this one is pretty involved. In order to show that $\overline{AC} \perp \overline{BD}$, you need to show that $\overline{AC}$ and $\overline{BD}$ intersect and form a right angle. So you need to show that $\angle AMB$ is a right angle. Because $\angle AMB$ and $\angle BMC$ form a straight angle, they are supplementary. If you could show that they were congruent, then they would have to be right angles. The easiest way to show $\angle AMB \cong \angle BMC$ is to show $\triangle AMB \cong \triangle CMB$ and use CPOCTAC. Now, because you're dealing with a rhombus, $\overline{AB} \cong \overline{BC}$. By the reflexive property of $\cong$, you know $\overline{BM} \cong \overline{BM}$. Because a rhombus is a parallelogram, and showed that the diagonals of a parallelogram bisect each other, $\overline{AM} \cong \overline{MC}$. You can use our SSS Postulate to conclude that $\triangle AMB \cong \triangle CMB$. Let's write it up.

Statements	Reasons
1. Parallelogram ABCD has diagonals $\overline{AC}$ and $\overline{BD}$ that intersect at M	Given
2. $\overline{AM} \cong \overline{MC}$	Theorem 15.6
3. $\overline{AB} \cong \overline{BC}$	Definition of a rhombus
4. $\overline{BM} \cong \overline{BM}$	Reflexive property of $\cong$
5. $\triangle AMB \cong \triangle CMB$	SSS Postulate
6. $\angle AMB \cong \angle BMC$	CPOCTAC

Statements	Reasons
7. ∠AMB and ∠BMC form a straight angle	Definition of straight angle
8. m∠AMB + m∠BMC = 180°	Angle Addition Postulate
9. m∠AMB + m∠AMB = 180°	Substitution (steps 6 and 8)
10. m∠AMB = 90°	Algebra
11. ∠AMB is right	Definition of right angle
12. $\overline{AC} \perp \overline{BD}$	Definition of ⊥

Squares

A square is both a rectangle and a rhombus. A *square* can be defined as a rectangle with congruent adjacent sides, or it could be defined as a rhombus that has a right angle. I'll pick the former description as the official definition. A square inherits all of the properties of a parallelogram, a rectangle, and a rhombus. A square has the best of all worlds. It has the properties of a parallelogram (opposite sides congruent, opposite angles congruent, opposite sides parallel, and diagonals bisect each other), a rectangle (diagonals are congruent and all four angles are congruent) and a rhombus (diagonals are perpendicular and all four sides are congruent).

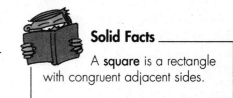

Solid Facts

A **square** is a rectangle with congruent adjacent sides.

Calculating Areas

Recall that in order to calculate the area of a triangle, we needed to know its height, or the length of its altitude (a line segment drawn perpendicularly from a vertex to the opposite side), and the length of its base (the side perpendicular to the altitude). The areas of trapezoids and parallelograms will also be calculated using the lengths of their altitudes and bases.

Area of Trapezoids

Recall that the two parallel sides of a trapezoid are its bases. The area A of a trapezoid whose bases have lengths b_1 and b_2 and whose altitude has length h is given by

$$A = \tfrac{1}{2}h(b_1 + b_2).$$

Figure 15.11 shows trapezoid ABCD with bases $\overline{BC}$ and $\overline{AD}$ and altitude $\overline{BE}$.

Figure 15.11

Trapezoid ABCD *whose bases have lengths* b_1 *and* b_2 *and whose altitude has length* h.

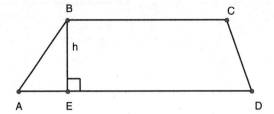

Example 5: Find the area of the trapezoid in Figure 15.11 if BC = 8, AD = 14, and BE = 5.

Solution: Because the area of a trapezoid is A = ½h(b_1 + b_2), you can substitute

our values for the height and the two bases. A = ½(5)(8 + 14), or A = 55 units².

Area of Parallelograms

There are a couple of methods for finding the area of a parallelogram, depending on the kind of parallelogram you are working with. The parallelogram shown in Figure 15.13 has an altitude of length h and a base of length b. Its area A can be calculated using the formula A = bh.

Figure 15.12

Parallelogram ABCD *has an altitude of length* h *and a base of length* b.

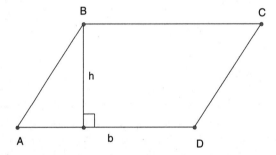

Example 6: Find the area of the parallelogram in Figure 15.12 if AD = 14 and BE = 5.

Solution: Using the equation for area, A = bh, A = 14 × 5 = 70 units².

This formula works with any parallelogram. If you are dealing with a rectangle, then the altitude corresponds to one of the sides of the rectangle and the base corresponds to one of its adjacent sides. So the area of a rectangle is the product of the lengths of two adjacent sides. If you are working with a square, then adjacent sides are congruent, so the area is the square of the length of one of its sides.

Whenever you are working with a quadrilateral whose diagonals are perpendicular, you can relate the area of the quadrilateral to the lengths of its diagonals. The quadrilateral ABCD in Figure 15.13 has diagonals of length d_1 and d_2. Its area can be calculated using the formula

$$A = \tfrac{1}{2}d_1d_2.$$

Eureka!

The area of a rectangle can be found by taking the product of the lengths of two consecutive sides. The area of a square is found by squaring the length of one of its sides.

This equation can be used for rhombuses and squares, and any other quadrilateral that has perpendicular diagonals, like kites.

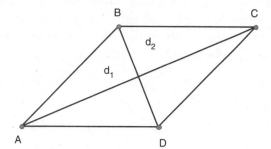

Figure 15.13

Quadrilateral ABCD *with diagonals of length* d_1 *and* d_2.

Example 7: Find the area of a rhombus ABCD with diagonals of length 8 and 12.

Solution: Substituting into the equation for area, $A = \tfrac{1}{2}d_1d_2$, you have

$A = \tfrac{1}{2}(8)(12)$, or $A = 48$ units².

The Pythagorean Theorem (again)

Now that you can compute the area of a square (it's just the square of the length of one of its sides), you can prove the Pythagorean Theorem again. Suppose we have a right triangle with legs of length a and b and the hypotenuse with length c as shown in Figure 15.14. Then by the Pythagorean Theorem, $a^2 + b^2 = c^2$.

Figure 15.14

A right triangle with legs of length a and b and the hypotenuse with length c.

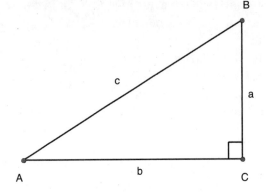

Arrange four copies of these triangles as shown in Figure 15.15. Then form two squares: a big square with side length a + b and a small square of side length c. The area of the big square can be calculated in two ways. The first way is to compute the square of the length of the side: $(a + b)^2$.

Figure 15.15

Four copies of △ABC arranged to help prove the Pythagorean Theorem.

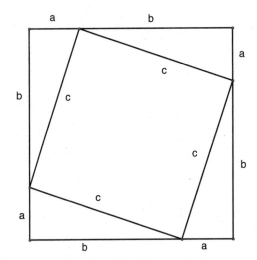

The second way is to add up the areas of the four triangles and the small square. The area of a right triangle is just one-half the product of the lengths of the legs:

½ab,

and the area of the small square is c^2. Because both of these methods for finding the area of the big square should give the same result, you have the following equation:

$(a + b)^2 = 4(½ab) + c^2$

Expanding your terms and rearranging yields the Pythagorean Theorem:

$$a^2 + 2ab + b^2 = 2ab + c^2$$

$$a^2 + b^2 = c^2$$

That's quite a slick proof of the Pythagorean Theorem, in my humble opinion. All you needed was a clever arrangement of four copies of the triangle and the ability to compute the areas of a triangle and a square.

Put Me in, Coach!

Here's your chance to shine. Remember that I am with you in spirit and have provided the answers to these questions in Appendix A.

1. In trapezoid ABCD, $\overline{BC} \parallel \overline{AD}$, R is the midpoint of $\overline{AB}$, and S is the midpoint of $\overline{CD}$, as shown in Figure 15.16. Find AD, BC, and RS if BC = 3x, RS = 4x + 9, and AD = 10x – 27.

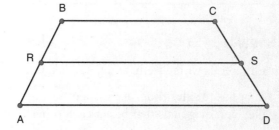

Figure 15.16

Trapezoid ABCD. $\overline{BC} \parallel \overline{AD}$, $\overline{AB}$ *has midpoint* R, *and* $\overline{CD}$ *has midpoint* S.

2. There are three other altitudes of trapezoid ABCD shown in Figure 15.4. Can you find them?

3. Write a proof of Theorem 15.3.

4. Write a proof of Theorem 15.4.

5. Write a two-column proof of Theorem 15.7.

6. Find the area of the trapezoid in Figure 15.11 if BC = 12, AD = 20 and BE = 9.

7. Find the area of the parallelogram in Figure 15.12 if AD = 20 and BE = 9.

8. Find the area of the parallelograms in Figure 15.17.

Figure 15.17

Compute the areas of the following parallelograms.

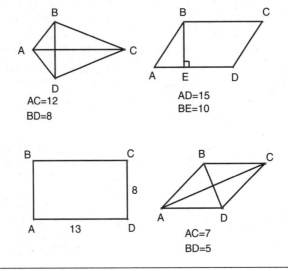

AC=12
BD=8

AD=15
BE=10

8

13

AC=7
BD=5

The Least You Need to Know

♦ Trapezoids are quadrilaterals with exactly two parallel sides.

♦ Parallelograms have opposite sides congruent, opposite angles congruent, opposite sides parallel, and diagonals that bisect each other.

♦ In addition to the properties of a parallelogram, a rectangle has congruent diagonals and all four angles are congruent.

♦ In addition to the properties of a parallelogram, a rhombus has congruent adjacent sides and perpendicular diagonals.

♦ In addition to the properties of a parallelogram, a square has the properties of a rectangle and a rhombus.

♦ The area of a trapezoid is $A = \frac{1}{2}h(b_1 + b_2)$; the area of a parallelogram is $A = bh$.

16

Proofs About Quadrilaterals

In This Chapter

- ◆ The "Name That Quadrilateral" game
- ◆ What makes a rectangle
- ◆ What makes a rhombus
- ◆ What makes a square

When the game *Twenty Questions* is played, one person thinks of an object, and the other people can ask "yes/no" questions to try and figure out what the object is. This game helps pass the time when traveling on long trips. On your next trip, if you find yourself bored with twenty questions and the license plate game, you might want to play a new game.

Now that you have our quadrilaterals in a row, it's time to learn how to pick them out of a line-up. In Chapter 15 you learned about the various properties of special quadrilaterals. You'll put that information to use by playing "Name That Quadrilateral." Here are the rules: I'll give you some clues about a quadrilateral, and you identify its type.

For example, I'm thinking of a parallelogram that has congruent and perpendicular diagonals. Name that parallelogram. To name that parallelogram, go down your list. Rectangles have congruent diagonals, but they aren't perpendicular. Rhombuses have perpendicular diagonals, but they aren't congruent. Squares, being a hybrid of rectangles and rhombuses, have congruent diagonals and perpendicular diagonals. We have a winner!

When Is a Quadrilateral a Parallelogram?

I'm thinking of a quadrilateral with one pair of opposite sides parallel and congruent. Name that quadrilateral.

I'm thinking of a quadrilateral with both pairs of opposite sides congruent. Name that quadrilateral.

I'm thinking of a quadrilateral with both pairs of opposite angles congruent. Name that quadrilateral.

I'm thinking of a quadrilateral whose diagonals bisect each other. Name that quadrilateral.

If you answered "parallelogram" to all of the above, you are correct! Of course, by now you know that it's not enough to claim that I'm thinking of a parallelogram. There are doubters in the car, so you will have to prove it.

Opposite Sides Congruent and Parallel

Your first "Name That Quadrilateral" clue involved one pair of opposite sides being parallel and congruent. I'll call it a theorem and write a two-column proof. Figure 16.1 will help you visualize the situation.

Figure 16.1

Quadrilateral ABCD *with* $\overline{BC} \parallel \overline{AD}$ *and* $\overline{BC} \cong \overline{AD}$.

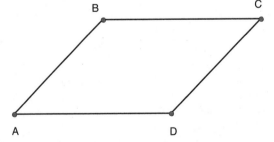

Theorem 16.1: If one pair of opposite sides of a quadrilateral are parallel and congruent, then the quadrilateral is a parallelogram.

Here's the game plan. Assume that $\overline{BC} \parallel \overline{AD}$ and $\overline{BC} \cong \overline{AD}$. By definition, a parallelogram is a quadrilateral with both pairs of opposite sides parallel. You already know that one pair of opposite sides is parallel. You need to show that the other pair of opposite sides is parallel. In other words, you need to show that $\overline{AB} \parallel \overline{CD}$.

You can look at this quadrilateral in two ways. The first way is to focus on segments $\overline{BC}$ and $\overline{AD}$ cut by a transversal $\overline{AC}$. Then $\angle BCA$ and $\angle DAC$ are alternate interior angles and are congruent because $\overline{BC} \| \overline{AD}$. The second way is to turn it on its side. $\overline{AB}$ and $\overline{CD}$ are two segments cut by a transversal $\overline{AC}$. In this case, $\angle BAC$ and $\angle ACD$ are alternate interior angles. If you could show that $\angle BAC \cong \angle ACD$, then you could conclude that $\overline{AB} \| \overline{CD}$, and you would be done. The way to show $\angle BAC \cong \angle ACD$ is to use CPOCTAC. In order to use CPOCTAC, you need to show $\triangle DAC \cong \triangle BCA$. In order to show $\triangle DAC \cong \triangle BCA$, you need to use the SAS Postulate. Let's write it up.

	Statements	Reasons
1.	Quadrilateral ABCD with $\overline{BC} \| \overline{AD}$ and $\overline{BC} \cong \overline{AD}$.	Given
2.	$\overline{BC} \| \overline{AD}$ cut by a transversal $\overline{AC}$	Definition of transversal
3.	$\angle BAC$ and $\angle ACD$ are alternate interior angles	Definition of alternate interior angles
4.	$\angle BCA \cong \angle DAC$	Theorem 10.2
5.	$\overline{AC} \cong \overline{AC}$	Reflexive property of $\cong$.
6.	$\triangle DAC \cong \triangle BCA$	SAS Postulate
7.	$\angle BAC \cong \angle ACD$	CPOCTAC
8.	$\overline{AB}$ and $\overline{CD}$ are two segments cut by a transversal $\overline{AC}$	Definition of transversal
9.	$\angle BAC$ and $\angle ACD$ are alternate interior angles	Definition of alternate interior angles
10.	$\overline{AB} \| \overline{CD}$	Theorem 10.8
11.	Quadrilateral ABCD is a parallelogram	Definition of parallelogram

Now that you have named that quadrilateral correctly, you can move on to the next quadrilateral.

Two Pairs of Congruent Sides

In the second "Name That Quadrilateral" game, the quadrilateral had two pairs of congruent sides. Let's write that as a theorem and lay it to rest.

> **Theorem 16.2**: If both pairs of opposite sides of a quadrilateral are congruent, then the quadrilateral is a parallelogram.

We have a visual in Figure 16.2. We have a parallelogram ABCD with $\overline{AB} \cong \overline{CD}$ and $\overline{BC} \cong \overline{AD}$. The game plan is to divide the quadrilateral into two triangles using the diagonal $\overline{AC}$. Use the SSS Postulate to show that the two triangles are congruent, and use CPOCTAC to conclude that alternate interior angles are congruent and opposite sides must be parallel. If we show this for both pairs of opposite sides, then we have a parallelogram by definition. It's time to write out the details.

Figure 16.2

Quadrilateral ABCD *with* $\overline{AB} \cong \overline{CD}$ *and* $\overline{BC} \cong \overline{AD}$.

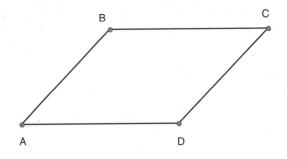

Statements	Reasons
1. Quadrilateral ABCD with $\overline{AB} \cong \overline{CD}$ and $\overline{BC} \cong \overline{AD}$	Given
2. $\overline{AC} \cong \overline{AC}$	Reflexive property of $\cong$
3. $\triangle ABC \cong \triangle CDA$	SSS Postulate
4. $\angle BAC \cong \angle ACD$ and $\angle BCA \cong \angle DAC$	CPOCTAC
5. $\overline{BC}$ and $\overline{AD}$ are two segments cut by a transversal $\overline{AC}$	Definition of transversal
6. $\angle BAC$ and $\angle ACD$ are alternate interior angles	Definition of alternate interior angles
7. $\overline{BC} \| \overline{AD}$	Theorem 10.8
8. $\overline{AB}$ and $\overline{CD}$ are two segments cut by a transversal $\overline{AC}$	Definition of transversal
9. $\angle BAC$ and $\angle ACD$ are alternate interior angles	Definition of alternate interior angles
10. $\overline{AB} \| \overline{CD}$	Theorem 10.8
11. Quadrilateral ABCD is a parallelogram	Definition of parallelogram

Once again, the sweet taste of victory! You have named that quadrilateral correctly. Next!

Two Pairs of Congruent Angles

The third description of the quadrilateral involved both pairs of opposite angles being congruent. I'll state the theorem and use Figure 16.3 to guide you through your proof.

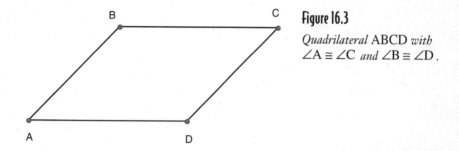

Figure 16.3

Quadrilateral ABCD *with* ∠A ≅ ∠C *and* ∠B ≅ ∠D.

Theorem 16.3: If both pairs of opposite angles of a quadrilateral are congruent, then the quadrilateral is a parallelogram.

You need to start with your angles. Because the measures of the sums of the interior angles of a quadrilateral add up to 360°, you can show $m\angle A + m\angle B = 180°$, or that ∠A and ∠B are supplementary angles. Now, you can look at this quadrilateral in the following light: $\overline{BC}$ and $\overline{AD}$ are two segments cut by a transversal $\overline{AB}$. Usually the transversal has been $\overline{AC}$, but this time you'll use $\overline{AB}$. Because your two angles on the same side of the transversal are supplementary, Theorem 10.10 tells you that $\overline{BC} \,||\, \overline{AD}$. A similar argument shows that $\overline{AB} \,||\, \overline{CD}$.

Statements	Reasons
1. Quadrilateral ABCD with ∠A ≅ ∠C and ∠B ≅ ∠D	Given
2. $m\angle A + m\angle B + m\angle C + m\angle D = 360°$	The measures of the interior angles of a quadrilateral add up to 360°
3. $m\angle A + m\angle B + m\angle A + m\angle B = 360°$	Substitution (steps 1 and 2)
4. $m\angle A + m\angle B = 180°$	Algebra
5. ∠A and ∠B are supplementary angles	Definition of supplementary angles
6. $\overline{BC}$ and $\overline{AD}$ are two segments cut by a transversal $\overline{AB}$	Definition of transversal
7. $\overline{BC} \,||\, \overline{AD}$	Theorem 10.10

continues

continued

Statements	Reasons
8.　$\overline{AB}$ and $\overline{CD}$ are two segments cut by a transversal $\overline{AD}$	Definition of transversal
9.　$m\angle A + m\angle D = 180°$	Substitution (steps 1 and 4)
10.　$\angle A$ and $\angle D$ are supplementary angles	Definition of supplementary angles
11.　$\overline{AB} \,\|\, \overline{CD}$	Theorem 10.10
12.　Quadrilateral ABCD is a parallelogram	Definition of parallelogram

Bisecting Diagonals

Ah, the last name game of this series! If you have a quadrilateral that has diagonals that bisect each other, your quadrilateral is a parallelogram. Figure 16.4 shows a parallelogram ABCD with diagonals $\overline{AC}$ and $\overline{BD}$ that intersect at M and bisect each other.

Figure 16.4

Quadrilateral ABCD *with diagonals* $\overline{AC}$ *and* $\overline{BD}$ *that intersect at* M *and bisect each other.*

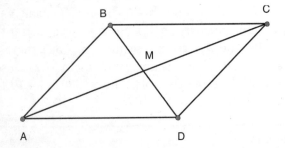

Theorem 16.4: If the diagonals of a quadrilateral bisect each other, then the quadrilateral is a parallelogram.

If you look at Figure 16.4, the game plan for proving this theorem should be coming through loud and clear. You will make use of Theorem 16.2: Pairs of opposite sides of a parallelogram are congruent. The two diagonals divide the parallelogram into four triangles. Because the diagonals bisect each other, $\overline{AM} \cong \overline{MC}$ and $\overline{BM} \cong \overline{MD}$. Because vertical angles are congruent, you can use the SAS Postulate to show that $\triangle AMD \cong \triangle BMC$ and $\triangle AMB \cong \triangle DMC$. From there it's a matter of applying CPOCTAC to show that both pairs of opposite sides are congruent.

Statements	Reasons
1. Quadrilateral ABCD with diagonals $\overline{AC}$ and $\overline{BD}$ that intersect at M and bisect each other	Given
2. $\overline{AM} \cong \overline{MC}$ and $\overline{BM} \cong \overline{MD}$	Definition of bisection
3. $\angle AMB \cong \angle CMD$ and $\angle AMD \cong \angle BMC$	Theorem 8.1
4. $\triangle AMD \cong \triangle BMC$ and $\triangle AMB \cong \triangle DMC$	SAS Postulate
5. $\overline{BC} \cong \overline{AD}$ and $\overline{AB} \cong \overline{CD}$	CPOCTAC
6. Quadrilateral ABCD is a parallelogram	Theorem 16.2

When Is a Parallelogram a Rectangle?

I'm thinking of a parallelogram whose diagonals are congruent. Name that parallelogram.

Not all parallelograms have congruent diagonals. Rhombuses do not have congruent diagonals. Rectangles do have congruent diagonals, and so do squares. You cannot conclude that the parallelogram that I'm thinking of is a square, though, because that would be too restrictive. When playing "Name That Quadrilateral," your answer must be as general as possible. Because a square is a rectangle but a rectangle need not be a square, the most general quadrilateral that fits this description is a rectangle.

> **Theorem 16.5**: If the diagonals of a parallelogram are congruent, then the parallelogram is a rectangle.

Figure 16.5 shows parallelogram ABCD with congruent diagonals $\overline{AC}$ and $\overline{BD}$. Because we are dealing with a parallelogram, you know that opposite sides are congruent. You can use the SSS Postulate to show that $\triangle ACD \cong \triangle DBA$. Using CPOCTAC, we can show $\angle A \cong \angle D$. Because ABCD is a parallelogram, opposite angles are congruent, so $\angle A \cong \angle C$ and $\angle B \cong \angle D$. By the transitive property of $\cong$, you have all four angles congruent. Because the measures of the interior angles of a quadrilateral add up to 360°, you can show that all four angles of our parallelogram are right angles. That's more than enough to make your parallelogram a rectangle.

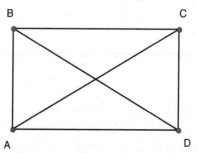

Figure 16.5

Parallelogram ABCD *with congruent diagonals* $\overline{AC}$ *and* $\overline{BD}$.

Statements	Reasons
1. Parallelogram ABCD with $\overline{AC} \cong \overline{BD}$	Given
2. $\overline{AB} \cong \overline{CD}$	Theorem 15.4
3. $\overline{AD} \cong AD$	Reflexive property of $\cong$
4. $\triangle ACD \cong \triangle DBA$	SSS Postulate
5. $\angle A \cong \angle D$	CPOCTAC
6. $\angle A \cong \angle C$ and $\angle B \cong \angle D$	Theorem 15.5
7. $m\angle A = m\angle C$ and $m\angle B = m\angle D$	Definition of $\cong$
8. $m\angle A + m\angle B + m\angle C + m\angle D = 360°$	The measures of the interior angles of a quadrilateral add up to 360°
9. $m\angle A + m\angle A + m\angle A + m\angle A = 360°$	Substitution (steps 7 and 8)
10. $m\angle A = 90°$	Algebra
11. $\angle A$ is a right angle	Definition of right angle
12. Parallelogram ABCD is a rectangle	Definition of rectangle

When Is a Parallelogram a Rhombus?

I'm thinking of a parallelogram whose diagonals are perpendicular. Name that parallelogram.

If you guessed that it was a square, then you didn't read the heading for this section very well. It's a rhombus! The nice thing about working with parallelograms is that the diagonals create lots of triangles just begging to be proven congruent. In Figure 16.6, parallelogram ABCD has perpendicular diagonals. The congruent triangles are trying to communicate with you. Listen closely.

Figure 16.6

Parallelogram ABCD *with* $\overline{AC} \perp \overline{BD}$.

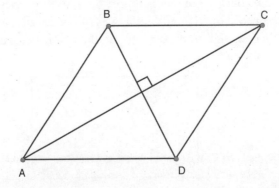

Theorem 16.6: If the diagonals of a parallelogram are perpendicular, the parallelogram is a rhombus.

Let's just jump right into the game plan. You know that $\overline{AC} \perp \overline{BD}$, so m∠AMB = 90° and m∠CMB = 90°. Because the diagonals of a parallelogram bisect each other, you know that $\overline{AM} \cong \overline{MC}$. The reflexive property of ≅ enables you to write $\overline{BM} \cong \overline{BM}$. By the SAS Postulate, you know that △AMB ≅ △CMB. By CPOCTAC, you know that $\overline{AB} \cong \overline{BC}$. Because $\overline{AB} \cong \overline{BC}$ and $\overline{AB} \cong \overline{BC}$ are adjacent sides, you have a parallelogram with congruent adjacent sides, a.k.a. a rhombus.

	Statements	Reasons
1.	Parallelogram ABCD has $\overline{AC} \perp \overline{BD}$	Given
2.	∠AMB and ∠CMB are right	Definition of ⊥
3.	m∠AMB = 90° and m∠CMB = 90°	Definition of right angle
4.	∠AMB ≅ ∠CMB	Definition of ≅
5.	$\overline{AM} \cong \overline{MC}$	Theorem 15.6
6.	$\overline{BM} \cong \overline{BM}$	Reflexive property of ≅
7.	△AMB ≅ △CMB	SAS Postulate
8.	$\overline{AB} \cong \overline{BC}$	CPOCTAC
9.	Parallelogram ABCD is a rhombus	Definition of rhombus

Now let's get a little tricky. Suppose you have a rectangle ABCD. Find the midpoints of each of the sides of the rectangle and join them together consecutively to form the quadrilateral MNOP, as shown in Figure 16.7. What kind of quadrilateral is made?

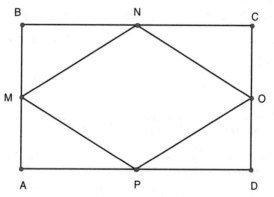

Figure 16.7

Rectangle ABCD, *with midpoints of each side joined together consecutively to form the quadrilateral* MNOP.

From the picture, it looks kind of like a parallelogram. You have to be careful, though, because looks can be deceiving. It also looks like the diagonals of the newly created quadrilateral are perpendicular. If the drawing is accurate, you might be tempted to conclude that the quadrilateral is a rhombus. Let's prove it.

Theorem 16.7: If the midpoints of the sides of a rectangle are joined in order, the quadrilateral formed is a rhombus.

You need a serious game plan for this one. Because M, N, O, and P are the midpoints of $\overline{AB}$, $\overline{BC}$, $\overline{CD}$, and $\overline{AD}$, you know that $\overline{BN} \cong \overline{NC} \cong \overline{AP} \cong \overline{PD}$ and $\overline{AM} \cong \overline{MB} \cong \overline{OD} \cong \overline{CO}$. Because you're dealing with a rectangle, you know that $m\angle A = 90°$, $m\angle B = 90°$, $m\angle C = 90°$, and $m\angle D = 90°$. So by the SAS Postulate, $\triangle PAM \cong \triangle NBM \cong \triangle PDO \cong \triangle NCO$. Applying the CPOCTAC principal $\overline{MN} \cong \overline{MP} \cong \overline{PO} \cong \overline{NO}$. So opposite sides are congruent and quadrilateral MNOP is a parallelogram. Also, adjacent sides are congruent, so parallelogram MNOP is a rhombus.

	Statements	Reasons
1.	Rectangle ABCD, with midpoints of each side joined together consecutively to form the quadrilateral MNOP	Given
2.	$\overline{BN} \cong \overline{NC}$, $\overline{AP} \cong \overline{PD}$, $\overline{AM} \cong \overline{MB}$ and $\overline{OD} \cong \overline{CO}$	Definition of midpoint
3.	$\overline{AB} \cong \overline{CD}$ and $\overline{BC} \cong \overline{AD}$	Theorem 15.4
4.	BN = ½BC, AP = ½AD, AM = ½AB and CO = ½CD	Theorem 9.1
5.	BN = NC = AP = PD and AM = MB = OD = CO	Substitution (steps 2, 3, 4)
6.	$\overline{BN} \cong \overline{NC} \cong \overline{AP} \cong \overline{PD}$ and $\overline{AM} \cong \overline{MB} \cong \overline{OD} \cong \overline{CO}$	Definition of $\cong$
7.	$m\angle A = 90°$, $m\angle B = 90°$, $m\angle C = 90°$ and $m\angle D = 90°$	Definition of rectangle
8.	$\triangle PAM \cong \triangle NBM \cong \triangle PDO \cong \triangle NCO$	SAS Postulate
9.	$\overline{MN} \cong \overline{MP} \cong \overline{PO} \cong \overline{NO}$	CPOCTAC
10.	Quadrilateral MNOP is a parallelogram	Theorem 16.2
11.	Quadrilateral MNOP is a rhombus	Definition of rhombus

When Is a Parallelogram a Square?

This is the last round of "Name That Quadrilateral." I'm thinking of a parallelogram with congruent perpendicular diagonals. Name that parallelogram.

Well, if a parallelogram has congruent diagonals, you know that it is a rectangle. If a parallelogram has perpendicular diagonals, you know it is a rhombus. So I'm thinking of a parallelogram that is both a rectangle and a rhombus. The only parallelogram that satisfies that description is a square.

Theorem 16.8: If the diagonals of a parallelogram are congruent and perpendicular, the parallelogram is a square.

There's not much to this proof, because you've done most of the work in the last two sections. The drawing in Figure 16.8 shows a parallelogram with congruent perpendicular diagonals, but it is misleading in that it does not quite look like a square. You won't be fooled by the picture, but you will extract the important information. Specifically, you need a parallelogram with diagonals $\overline{AC}$ and $\overline{BD}$ that are both perpendicular and congruent. The reason I intentionally drew a generic parallelogram rather than a square is that I want to be careful not to assume what I am trying to prove. If I drew a square, I might be tempted to draw conclusions about the lengths of the adjacent sides. As mathematicians in training, it is important that you stay as far away from the pitfall of assuming what you are trying to prove. It's not uncommon for people to enjoy the scenery on their geometric trip and forget to watch where they walk!

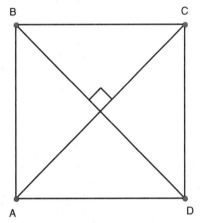

Figure 16.8

Parallelogram ABCD *with* $\overline{AC} \cong \overline{BD}$ *and* $\overline{AC} \perp \overline{BD}$.

	Statements	Reasons
1.	Parallelogram ABCD with $\overline{AC} \cong \overline{BD}$ and $\overline{AC} \perp \overline{BD}$	Given
2.	Parallelogram ABCD is a rectangle	Theorem 16.5
3.	Parallelogram ABCD is a rhombus	Theorem 16.6
4.	Parallelogram ABCD is a rectangle with congruent adjacent sides	Definition of a rhombus
5.	Parallelogram ABCD is a square	Definition of a square

That wraps up this episode of "Name That Quadrilateral." Contestants earned a solid education, which can be passed on to friends, family, and anyone willing to listen. If you plan to be in the area and would like tickets to the taping of "Name That Quadrilateral," send a postcard to the address at the end of the book with your name, grade level, mathematical interests, and dates that you'll be in the area. See you next time!

The Least You Need to Know

◆ There are four ways to recognize a parallelogram: one pair of opposite sides congruent and parallel, both pairs of opposite sides are congruent, both pairs of opposite angles are congruent, or the diagonals bisect each other.

◆ If the diagonals of a parallelogram are congruent, you actually have a rectangle.

◆ If the diagonals of a parallelogram are perpendicular, you actually have a rhombus.

◆ If the diagonals of a parallelogram are congruent and perpendicular, you actually have a square.

Part 4

Going Around in Circles

You've had the experience of spinning your wheels. Maybe you've ridden a bicycle, a skateboard, or a scooter. You must have ridden in a car. There are times when spinning your wheels has gotten you somewhere and you've successfully traveled from point A to point B. There are other times when your wheels have sunk into mud, sand, or dirt, or they can't get a grip on the ice. In this case, your frustration level rises and you can't go anywhere.

Going around in circles can be comforting (like when you go to school and then return home) or it can be irritating (like when you get a bureaucratic runaround). In geometry, circles can help you get a handle on things like angle measures and segment lengths. They can even help you prove things about triangles!

Chapter 17

Anatomy of a Circle

In This Chapter

◆ Dissect a circle

◆ How to make chords

◆ Go off on several tangents

◆ Arc length and area

If you have ever tried to get information from a bureaucratic organization, you have probably gotten the "run around." The phrase "going around in circles" doesn't usually conjure up an image of progress. But if you think about the industrial revolution, "going around in circles" was exactly how progress was made. Pulleys and gears enabled machines to increase production while unions helped shorten the workweek.

Circles and cycles go hand in hand. Cycles are made by going around a circle. You study the cycles of the earth for many reasons: to help you decide which clothes to pack for vacation, when you should plant your garden, and what time you should get up to see sunrise.

Without an understanding of circles, it would be difficult to talk about the circle of life, explain why you do not fall off the edge of the earth, or give change for a dollar. Circles are all around us, and it is worthwhile to understand them as much as possible.

Basic Terms

Start with a point C, and collect all of the points a fixed distance r units away from it. Give this collection of points a name: *circle*. I have drawn a circle in Figure 17.1. The starting point is called the *center* of the circle. Any line segment having the center of the circle as one endpoint and any point on the circle as the other endpoint is called a *radius* of the circle. Because all points on the circle are a distance of r units away from the center, all radii of a circle are congruent. This will be stated as a theorem, though the proof would take no more than a line or two, with the reasons being either "given" or "definition of a circle."

Figure 17.1

A circle with center C *and radius* $\overline{AC}$.

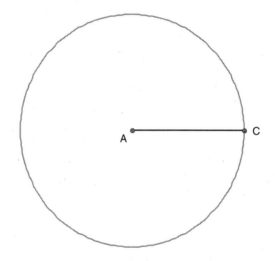

Theorem 17.1: All radii of a circle are congruent.

Solid Facts _____

A **circle** is the set of all points in a plane that are a fixed distance from a given point.

The **center** of the circle is the point equidistant from all points on the circle.

A **radius** of a circle is a line segment with one endpoint being the center of the circle, the other endpoint being a point on the circle.

Don't limit yourself to only drawing radii of circles. Circles get really interesting when you connect points on a circle. A line segment that joins two points on the circle is called a *chord* of the circle. A *diameter* of a circle is a chord that contains the center of the circle. The length of a diameter of a circle is twice the length of the radius of a circle. This can be proven by using the Segment Addition Postulate (Postulate 3.5).

If you have three points on a circle, you can connect them to form an inscribed angle. An *inscribed angle* of a circle is an angle whose vertex is a point on the circle and whose sides are chords of the circle. You can also construct *inscribed polygons* by using points on the circle as the vertices.

Two circles that coincide are congruent. In order for two circles to be congruent, the lengths of the radii must be congruent. If two circles have the same center they are called *concentric circles*.

There are times when you will need to measure the distance around the circle. When you did this with polygons, it was called the perimeter. With circles, it will be called the *circumference* of the circle.

Solid Facts _____

A **chord** is a line segment that joins two points on a circle.

A **diameter** of a circle is a chord that contains the center of the circle.

An **inscribed angle** of a circle is an angle whose vertex is a point on the circle and whose sides are chords of the circle.

An **inscribed polygon** of a circle is a polygon whose vertices are points on the circle and whose sides are chords of the circle.

Congruent circles are circles that have congruent radii.

Concentric circles are circles that have the same center.

The **circumference** of a circle is the linear measure of the distance around the circle.

Arcs

I know that I've just thrown a lot of new terminology at you, but I'm not done. I've connected points on a circle with straight line segments. It is also possible to connect points on a circle using the curvy part of the circle. Suppose you have two points, A and B, on the circle, as shown in Figure 17.3. The points between A and B form the line segment $\overline{AB}$, and the points between A and B that lie on the circle make up the *arc* AB. Because the arc AB is curvy (it consists of part of the circle), it is given the abbreviation $\overparen{AB}$.

Figure 17.2

$\overline{AB}$ *and* $\overset{\frown}{AB}$ *on a circle.*

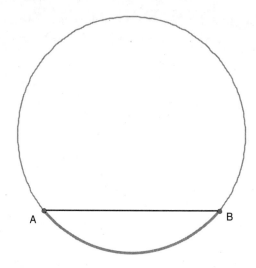

There's just one problem with this idea. There are actually *two* arcs that include the points on the circle between A and B, and you need to be able to distinguish between the two. You can do this by specifying a point in between A and B on the correct piece of the circle. So the upper arc is $\overset{\frown}{ADB}$, and the lower arc is $\overset{\frown}{AEB}$. If A and B are endpoints of a diameter, $\overset{\frown}{ADB}$ and $\overset{\frown}{AEB}$ are called semicircles. An arc of a circle is either a semicircle, part of a semicircle (called a minor arc), or more than a semicircle but less than an entire circle (called a major arc).

Eureka! _____

The difference between $\overline{AB}$ and $\overset{\frown}{AB}$ is that $\overset{\frown}{AB}$ is the part of the circle that connects A and B, and $\overline{AB}$ is the straight line segment that connects A and B.

Tangent Line

If two arcs are part of two circles with different radii, then there is no way that they can have the same curvature (or be congruent) The only way for two arcs to be congruent is if the circles they are on are congruent and their arc lengths are congruent.

Because you can measure the lengths of line segments, it is natural to ask about measuring arcs. Arcs have length and curve. To measure the length of an arc, imagine straightening the arc and measuring its length with your ruler. The two arcs on the left in Figure 17.3 have the same length but not the same curvature. The arcs on the right have the same curvature but not the same length. So when you measure arc length, there are two features you need to address: actual length and curvature. In a circle (or in congruent circles), *congruent arcs* are arcs that match up both in length and in curvature.

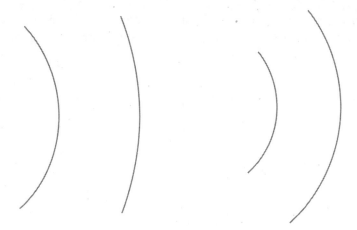

Figure 17.3

Left: Two arcs with the same length but different curve. Right: Two arcs with the same curve, but different length.

The curvature of an arc is measured in degrees. The measure of $\overarc{AB}$ will be denoted $m\overarc{AB}$. In order to measure an arc, I need to introduce you to a central angle. A central angle of a circle is an angle whose vertex is the center of the circle and whose sides are radii of the circle. The reason that central angles are needed when measuring the curvature of an arc $\overarc{AB}$ can be seen in Figure 17.4. The minor arc AB and the central angle $\angle ACB$ are related to each other. Every arc has an associated central angle, and every central angle has an associated arc. If you are given a central angle, its *intercepted arc* is determined by the two points of intersection of the angle with the circle and all points of the arc in the interior of the angle.

Figure 17.4

A central angle and its corresponding minor arc.

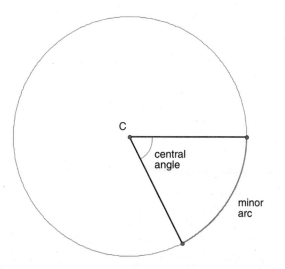

The correspondence between central angles and intercepted arcs can be used to define the curvature of an arc (also known as its degree measure). To solidify this idea, you will need a postulate.

> **Postulate 17.1:** Central Angle Postulate. In a circle, the degree measure of a central angle is equal to the degree measure of its intercepted arc.

This postulate enables you to determine the degree measurement of an arc by determining the measurement of its central angle. Because you have already explored angles in great detail, you have laid the foundation for understanding the degree measurement of an arc. A minor arc has a corresponding central angle whose measure is less than 180°; a semicircle has a central angle that is straight, so its degree measure equals 180°; and a major arc has a degree measure greater than 180°. Because the notion of degree measurement of an arc and central angle measurement are combined, central angle measurements greater than 180° are allowed. And if you throw in the Angle Addition Postulate (Postulate 4.2), you see that the degree measurement of the entire circle is twice the degree measurement of a semicircle. In other words, the degree measurement of a circle is 360°. Figure 17.5 shows some central angles and their intercepted arcs.

Figure 17.5

Central angles and their intercepted arcs.

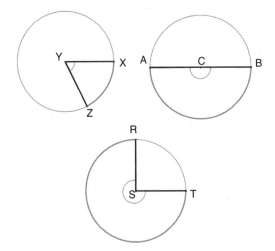

The last term I need to define for you is a sector of a circle. Imagine that you have ordered a slice of pizza. The two straight edges of the pizza can be viewed as radii of the circle (assuming it was cut properly), and the crusty part of the pizza is the arc of the circle. The entire pizza slice is a sector of the circle. A *sector of a circle* is the region bounded by two radii of the circle and the intercepted arc.

Solid Facts _____

An **arc** is the part of a circle determined by two points and all points between them.

Congruent arcs are arcs on circles with congruent radii that have the same degree measure.

A **minor arc** is an arc whose degree measure is between 0° and 180°.

A **semicircle** is an arc whose degree measure is exactly 180°.

A **major arc** is an arc whose degree measure is between 180° and 360°.

A **sector of a circle** is the region bounded by two radii of the circle and the intercepted arc.

You have already seen the Segment Addition Postulate and the Angle Addition Postulate. There is also an Arc Addition Postulate, which says exactly what you would expect it to say.

> **Postulate 17.2:** Arc Addition Postulate. If B lies between A and C on a circle, then $m\widehat{AB} + m\widehat{BC} = m\widehat{AC}$.

You will use this postulate when you combine arcs of circles.

Circumference and Area: Pi Anyone?

One of the unusual things about circles is that they are all similar to each other. I mean this in the official sense. It's not just that circles have the same shape but that everything is proportional as well. It is from proportionalities that the following postulate is presented.

> **Postulate 17.3:** The ratio of the circumference of a circle to the length of its diameter is a unique positive constant.

This constant of proportionality is represented by the Greek letter π (pronounced Pi, as in key lime). The reason that a Greek letter is used is because π is an irrational number. It cannot be written as neatly as 3 or 3.1, or 22/7, even though these numbers are approximations to π. It is impossible to write down the precise value of π.

If you denote the circumference of a circle by C and the length of a diameter of the circle by d, then C = πd. Because the length of a diameter of a circle is twice the length of the radius, if you let r denote the length of the radius of the circle, you have d = 2r and C = 2πr.

Tangent Line

Real numbers are classified as follows: the positive, whole numbers are called the natural numbers. The positive and negative whole numbers make up the integers. The collection of the ratios of integers comprises the rational numbers. The real numbers that are not rational numbers are called irrational numbers. An irrational number cannot be written as the ratio of two integers. The Greeks knew that irrational numbers (like $\sqrt{2}$ and π) existed, and they were able to estimate π very accurately.

Now that you know how to compute the circumference of a circle, it is time to revisit arcs. Remember that there are two measurements associated with arcs: degree measurement and length. It turns out that the lengths of arcs can be found using a proportionality. The ratio of the degree measure m of an arc to 360° (the degree measure of a circle) is the same as the ratio of the length l of the arc to the circumference. The equation is given by

$$\frac{m}{360°} = \frac{l}{C} = \frac{l}{2\pi r}.$$

Rearranging this equation to solve for the arc length, you get

$$l = \frac{m}{360°}C.$$

Let's do some calculations.

> **Example 1**: Find the circumference of the circle and the length of $\overset{\frown}{ABC}$ in Figure 17.6.

Figure 17.6

The circle with radius 5 inches, and $m\angle AOC = 125°$.

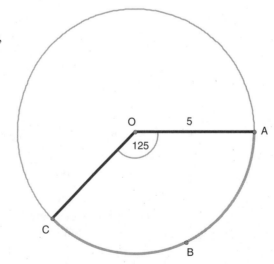

Solution: To find the circumference of the circle, just use the equation that relates the circumference of a circle to the radius of the circle:

$C = 2\pi r = 2\pi(5) = 10\pi$ inches.

To find the length of $\overset{\frown}{ABC}$, notice that the central angle has measure 125°. Using the equation for arc length, you have:

$$l = \frac{m}{360°}C = \frac{125°}{360°}(10\pi) = \frac{125\pi}{36} \text{ inches}$$

Now consider the problem of finding the area A of a circle whose radius has length r. The formula for this is $A = \pi r^2$. You can find the area of a sector using proportions, just as you found the length of an arc. If the area of the entire circle is πr^2, the area of a sector with degree measure θ is found by evaluating

$$A_{\text{sector}} = \frac{\theta\pi r^2}{360}.$$

Example 2: Find the area of a circle whose radius has a length of 15 inches.

Solution: You can solve this problem easily using the equation for area:

$A = \pi r^2 = \pi(15)^2 = 225\pi$ inches²

Example 3: Find the area of the sector with a radius length of 4 inches and central angle measuring 120°.

Solution: Substituting into our equation, you see that the area is

$$\frac{\theta\pi r^2}{360} = \frac{120\pi(4)^2}{360} = \frac{16\pi}{3} \text{ inches}^2.$$

Tangents

You might have observed that I occasionally go off on a tangent. In fact, I've gone off on several tangents just in this chapter alone! When someone goes off on a tangent, it's usually because some memory or thought is triggered. It usually involves going off the subject, in some completely different direction, and the audience is left wondering how the current topic of conversation relates to the one they just came from. The phrase "going off on a tangent" has its basis in geometry. A tangent is a line that intersects a circle at exactly one point. The point of intersection is the point of tangency.

Figure 17.7 shows a circle with a tangent line. Notice that the line only intersects the circle once and seems to glance off of the circle as it travels on its way. If you draw the radius of the circle that passes through the point of tangency, you will see that that radius and the tangent line are perpendicular.

Figure 17.7

A line tangent to a circle.

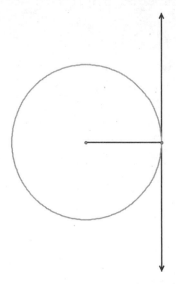

Tangent Line

Tangent lines have their applications in physics as well. If you put a stone in a sling and spin it around, faster and faster, you are spinning the stone and the sling in the shape of a circle. If you then release the stone, it will travel in a direction tangent to the spinning circle.

From One Theorem Comes Many

You haven't had a proof in awhile. "Never let a chapter go by without a proof," I always say. Here's an interesting tidbit about inscribed angles that might surprise you.

> **Theorem 17.2**: The measure of an inscribed angle of a circle is one-half the measure of its intercepted arc.

There are three ways to create an inscribed angle: either one of the sides of the angle is a diameter, each side of the angle lies on different sides of a diameter, or both sides of the angle lie on the same side of a diameter. I have illustrated each of these situations in Figure 17.8.

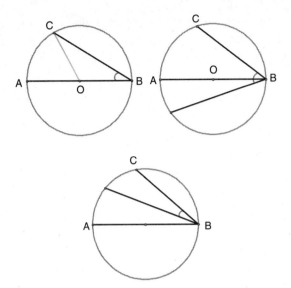

Figure 17.8

An inscribed circle where (a) one side of the angle is a diameter; (b) each side of the angle lies on a different side of the diameter $\overline{AB}$, and (c) both sides of the angle lie on the same side of the diameter $\overline{AB}$.

The proof of the first case requires the most detail. The second and third cases use the first case and the Arc Addition and Angle Addition Postulates.

After you prove this theorem, many other theorems will follow. Their proofs will be very short, because most of the work will have been done here.

Given: You are given a circle with center O, diameter $\overline{AB}$, and inscribed angle $\angle ABC$. The goal is to prove that $m\angle ABC$ is equal to one-half of the degree measure of $\overset{\frown}{AC}$. This situation is shown in Figure 17.8(a).

Proof: The game plan for this proof involves creating a central angle, because the only way you know to determine the degree measure of an arc is to relate it to the measure of its corresponding central angle. Suppose $\angle ABC$ is inscribed in a circle with center O, as shown in Figure 17.8(a). The goal is to relate that degree measure of $\overset{\frown}{CA}$ and $m\angle CBA$. Let's make a triangle.

You might be tempted to make $\triangle ABC$, but then you would have two of the sides being chords, and one side being a diameter. You wouldn't have enough useful information to draw any conclusions. The other triangle you can make is $\triangle COB$. This triangle is useful for two reasons. The first reason is that you have a central angle, $\angle COA$, which will enable you to find the degree measure of $\overset{\frown}{CA}$ (the arc you want to relate to your angle). The second reason that this triangle is useful is that two of its three sides are radii of the circle. Because all radii of a circle are congruent, you will have created an isosceles triangle. Isosceles triangles have the property that the angles opposite the congruent sides are congruent:

∠OCB ≅ ∠OBC . Because ∠COA is an exterior angle to ΔCOB, you know that m∠COA = m∠OCB + m∠OBC . Because you know the degree measure of $\overset{\frown}{CA}$ is equal to m∠COA, the desired relationship will be established. Let's organize this proof into its columns.

Statements	Reasons
1. A circle with center O, diameter $\overline{AB}$, and inscribed angle ∠ABC	Given
2. Construct the radius $\overline{CO}$	Postulate 3.2
3. $\overline{CO} ≅ \overline{OB}$	Theorem 17.1
4. ΔCOB is an isosceles triangle	Definition of isosceles triangle
5. ∠OCB ≅ ∠OBC	Theorem 12.1
6. m∠OCB = m∠OBC	Definition of ≅
7. ∠COA is an exterior angle to ΔCOB	Definition of exterior angle
8. m∠COA = m∠OCB + m∠OBC	Theorem 11.3
9. m∠COA = 2m∠OBC	Substitution (steps 6 and 8)
10. m∠OBC = ½m∠COA	Algebra
11. ∠COA is a central angle	Definition of central angle
12. The degree measure of $\overset{\frown}{AC}$ is equal to m∠COA	Central Angle Postulate
13. m∠OBC is equal to one-half the degree measure of $\overset{\frown}{AC}$	Substitution (steps 10 and 12)

Now that the proof of Theorem 17.2 is out of the way, you can prove several other theorems and calculate some arc and angle measures.

Theorem 17.3: An angle inscribed in a semicircle is a right angle.

This situation is illustrated in Figure 17.9. Because an inscribed angle has a measure equal to one-half the measure of the intercepted arc, and a semicircle has degree measure equal to 180°, you know that the inscribed angle has measure equal to one-half of 180°, which happens to equal 90°, which means that the inscribed angle is a right angle. Nothing too earth-shattering there!

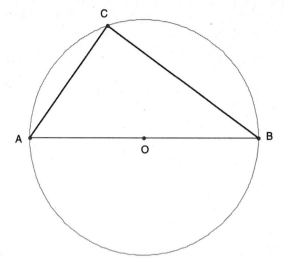

Figure 17.9

An angle inscribed in a semicircle.

Theorem 17.4: If two inscribed angles intercept the same arc, then these angles are congruent.

Again, each inscribed angle has measure equal to one-half of the measure of the intercepted arc. Look at Figure 17.10. Because the same arc is intercepted, each inscribed angle has the same measure, so the two inscribed angles are congruent. These theorems seem to fall out naturally from Theorem 17.2!

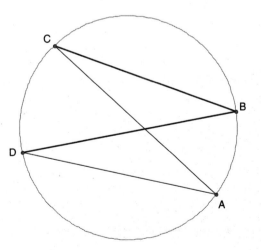

Figure 17.10

Two inscribed angles, ∠CBD and ∠CAD, intercept the same arc.

Example 4: If, in Figure 17.11, m∠CBD = 34°, find the degree measure of the minor arc $\overset{\frown}{CD}$.

Solution: Because the measure of the inscribed angle is one-half the degree measure of the intercepted arc (or the degree measure of the intercepted arc is twice the measure of the inscribed angle), the degree measure of $\overarc{CD}$ is twice m$\angle$CBD, or 68°.

Put Me in, Coach!

Here's your chance to shine. Remember that I am with you in spirit and have provided the answers to these questions in Appendix A.

1. Find the circumference of the circle and the length of $\overarc{RST}$ in Figure 17.11.

Figure 17.11

The circle with radius length 10 feet, and
m$\angle$ROT = 155°.

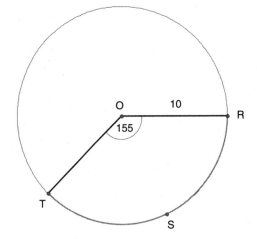

2. Find the area of a circle whose radius has a length of 3 feet.

3. Find the area of a sector with a radius of length 6 inches and a central angle measuring 150°.

4. If, in Figure 17.10, the degree measure of the minor arc $\overarc{CD}$ is 56°, find m$\angle$CBD.

The Least You Need to Know

- In a circle, the degree measure of a central angle is equal to the degree measure of its intercepted arc.

- Pi (π) is the ratio between the circumference of a circle and the length of a diameter of that circle.

- The circumference of a circle can be calculated using the formula $C = 2\pi r$, and the area of a circle can be found using the equation $A = \pi r^2$.

- The measure of an inscribed angle is one-half the degree measure of its intercepted arc.

Chapter 18

Segments and Angles

In This Chapter

◆ Using triangles in your proofs about circles

◆ Angles and chords

◆ Relating chords and arcs

◆ Calculating areas

Experienced pizza makers know all about cutting up circles. Their knives carve chords into the pie. An experienced slicer will help prevent food fights by dividing the pies along diameters so that the pizza slices are all congruent.

Some pizza eaters do not like the crust. They will eat a triangle of pizza and leave the sliver of crust. In order to determine your caloric intake (and whether or not enough pizza was eaten in order to justify getting desert), it might be useful to find the area of the pizza that was actually eaten. We can even find the area of the sliver of crust that was left behind. After you have read this chapter, you will be amazed at the calculations you can do with your leftovers.

There are relationships between angles and chords, arcs and chords, and radii and chords. It's time to dive right in and let the proofs begin!

Angles and Chords

Suppose a circle has two chords, as shown in Figure 18.1. If these two chords are congruent, it turns out that the central angles are congruent. Not only that, but if the two central angles are congruent, it turns out that the chords are congruent. Strange but true. And after you write out the proofs, it will no longer be strange. Just true.

Figure 18.1

A circle with two chords.

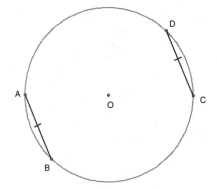

Theorem 18.1: If two chords of a circle are congruent, then the corresponding central angles are congruent.

Theorem 18.2: If two central angles are congruent, then the corresponding chords are congruent.

I'll write the proof for Theorem 18.1. I'll let you solo on the proof of Theorem 18.2.

Example 1: Prove Theorem 18.1.

Solution: Figure 18.2 will help sort through this mess. It shows two chords, $\overline{AB}$ and $\overline{CD}$, as well as their corresponding central angles.

Figure 18.2

A circle with two congruent chords: $\overline{AB} \cong \overline{CD}$.

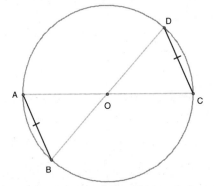

Given: $\overline{AB} \cong \overline{CD}$.

Prove: $\angle BOA \cong \angle COD$.

Proof: Here's the game plan. In trying to prove that $\angle BOA \cong \angle COD$, create the triangles $\triangle AOB$ and $\triangle COD$. Two of the sides of each of these triangles are radii, and hence are congruent. You are given that the other two sides of the triangles are congruent, so you can make use of the SSS Postulate (Postulate 12.1). Use CPOCTAC to conclude that $\angle BOA \cong \angle COD$.

Statements	Reasons
1. A circle with center O and chords $\overline{AB}$ and $\overline{CD}$, with $\overline{AB} \cong \overline{CD}$	Given
2. $\overline{AO}, \overline{BO}, \overline{CO}$, and $\overline{DO}$ are radii	Definition of radii
3. $\overline{AO} \cong \overline{CO}$ and $\overline{BO} \cong \overline{DO}$	Theorem 17.1
4. $\triangle AOB \cong \triangle COD$	SSS Postulate
5. $\angle BOA \cong \angle COD$	CPOCTAC

When you prove Theorem 18.2 you'll still want to use CPOCTAC, but in proving the triangles are congruent you'll need to use the SAS Postulate (Postulate 12.2).

Arcs and Chords

Now that you know the connection between angles and chords, it doesn't take a genius to figure out the corresponding relationship between arcs and chords. Because central angles and intercepted arcs have the same degree measure, you can take the results of the previous section and make one minor adjustment: Replace "central angle" with "intercepted arc."

Theorem 18.3: If two chords of a circle are congruent, then the intercepted arcs are congruent.

Theorem 18.4: If two arcs are congruent, then the corresponding chords are congruent.

I really should leave both of these proofs for you, but then it might appear that I am not doing my fair share. So I'll take the first theorem and you can take the second one.

Example 2: Prove Theorem 18.3.

Solution: Figure 18.3 shows what is going on. I have two congruent chords, $\overline{AB}$ and $\overline{CD}$, which have intercepted arcs $\overparen{AEB}$ and $\overparen{CFD}$.

Figure 18.3

A circle with two congruent chords: $\overline{AB} \cong \overline{CD}$.

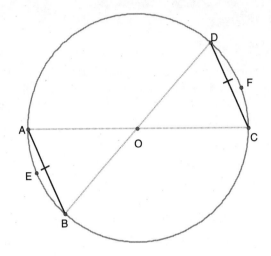

Given: $\overline{AB} \cong \overline{CD}$.

Prove: $\widehat{AEB} \cong \widehat{CFD}$.

Proof: The game plan is to use Theorem 18.1, which has just been proven, to show that the central angles are congruent. Because the degree measure of the central angle is equal to the degree measure of the intercepted arc (this is just the Central Angle Postulate, Postulate 17.1), you can stick a fork in this theorem because it is done!

Statements	Reasons
1. A circle with center O and chords $\overline{AB}$ and $\overline{CD}$, with $\overline{AB} \cong \overline{CD}$	Given
2. $\angle BOA$ and $\angle COD$ are central angles	Definition of central angles
3. $\angle BOA \cong \angle COD$	Theorem 18.1
4. $m\angle BOA = m\angle COD$	Definition of $\cong$
5. $m\angle BOA = m\widehat{AEB}$ and $m\angle COD = m\widehat{CFD}$	Central Angle Postulate
6. $m\widehat{AEB} = m\widehat{CFD}$	Substitution (steps 4 and 5)
7. $\widehat{AEB} \cong \widehat{CFD}$	Definition of $\cong$

When you prove Theorem 18.4, you'll probably want to make use of Theorem 18.2. But that's just a hunch.

Radii and Chords

Now I think that you are ready for some real fun. Suppose you have a chord $\overline{AB}$ (which doesn't necessarily pass through the center of the circle), and a radius $\overline{OC}$ of the circle that is perpendicular to $\overline{AB}$, as shown in Figure 18.4. Call the point of intersection of $\overline{OC}$ and $\overline{AB}$ point D. What kinds of things can you prove?

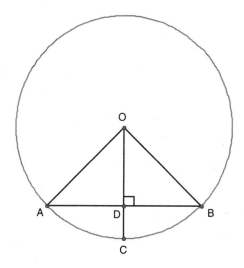

Figure 18.4

A circle with center O, chord $\overline{AB}$, and radius $\overline{OC}$ with $\overline{OC} \perp \overline{AB}$, and $\overline{AB}$ intersecting $\overline{OC}$ at D.

$\overline{OA}$ and $\overline{OB}$ are radii, so they are congruent. Because $\overline{OC} \perp \overline{AB}$, $\angle ODA$ and $\angle ODB$ are right angles, and $\triangle ODA$ and $\triangle ODB$ are right triangles. By the HL Theorem, these two triangles are congruent. You can make use of CPOCTAC to show lots of angle and segment relationships. I'll specify a few.

> **Theorem 18.5**: If a radius is perpendicular to a chord, it bisects that chord.

> **Theorem 18.6**: If a radius is perpendicular to a chord, it bisects the arc of that chord.

I will prove the first theorem and let you prove the second theorem.

> **Example 3**: Prove Theorem 18.5.

> **Solution**: Figure 18.4 is our starting point. In terms of the drawing, you are given a circle with center O, chord $\overline{AB}$, and radius $\overline{OC}$ with $\overline{AB}$ intersecting $\overline{OC}$ at D. Furthermore, $\overline{OC} \perp \overline{AB}$. The goal is to prove that $\overline{OC}$ bisects $\overline{AB}$.

Proof: The game plan has already been discussed. The only thing I haven't mentioned is that you'll need to use the definition of bisector (meaning you'll need to show $\overline{AD} \cong \overline{DB}$). But I doubt that surprises you.

	Statements	Reasons
1.	A circle with center O, chord $\overline{AB}$, and radius $\overline{OC}$ with $\overline{AB}$ intersecting $\overline{OC}$ at D and $\overline{OC} \perp \overline{AB}$	Given
2.	$\angle ODA$ and $\angle ODB$ are right angles	Definition of $\perp$
3.	$\triangle ODA$ and $\triangle ODB$ are right triangles	Definition of right triangles
4.	$\overline{OA}$ and $\overline{OB}$ are radii of the circle	Definition of radii
5.	$\overline{OA} \cong \overline{OB}$	Theorem 17.1
6.	$\overline{OD} \cong \overline{OD}$	Reflexive property of $\cong$
7.	$\triangle ODA \cong \triangle ODB$	HL Theorem for right triangles
8.	$\overline{AD} \cong \overline{DB}$	CPOCTAC
9.	$\overline{OC}$ bisects $\overline{AB}$	Definition of bisect

The proof of Theorem 18.6 will just take things one step further. Using CPOCTAC, you can show that two central angles are congruent and then use the Central Angle Postulate.

There's one more theorem to prove in this chapter. You might be disappointed to learn that I'm out of theorems, and it might seem a bit unfair that I get to write the last proof of the chapter. But there are a few applications of this theorem to work out, so at least you'll get some practice using it.

This last theorem is related to Theorem 18.5. In this situation, you have a chord and a line that is perpendicular to your chord and bisects it, as shown in Figure 18.5. It turns out that that is enough to guarantee that the line passes through the center of your circle.

Figure 18.5

A circle, a chord, and its perpendicular bisector.

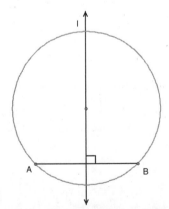

You might be wondering why this result isn't trivial. Well, there are lots of lines that can bisect a line segment. I've shown two of them in Figure 18.6. If M is the midpoint of $\overline{AB}$, lines l and m both bisect $\overline{AB}$ because they both pass through M. All it takes to bisect a segment is to cut it in half. Because both lines pass through the midpoint, both lines cut your segment in half. Because there are two distinct lines that intersect at M, they can't intersect anywhere else. Now imagine that this segment is a chord of a circle. You cannot have both lines pass through the center of your circle (otherwise your lines would intersect twice—once at M and once at the center—and that is not allowed!).

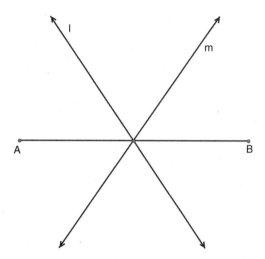

Figure 18.6

$\overline{AB}$ *has midpoint* M *and two distinct lines,* l *and* m, *which pass through* M, *bisecting* $\overline{AB}$.

So, something else is needed in order to ensure that the line passes through the center of our circle. Bisecting your chord is not enough. Look back at Theorem 18.5. What else did you know about the radius and the chord? You assumed that they were perpendicular! And that's the other piece needed here.

Theorem 18.7: The perpendicular bisector of a chord contains the center of the circle.

The objective is to show that the perpendicular bisector passes through the origin. The best way to tackle this problem is to use contradiction. Suppose that your line is the perpendicular bisector of a chord with midpoint M and it does *not* pass through the center of the circle. Let's explore what kind of havoc will result. Create the radius of our circle that *does* pass through M and call it $\overline{OD}$. I've drawn this for you in Figure 18.7. This radius cannot lie on our line because our line does not pass through the center. So the line and the radius intersect (at M) and form an angle. There's always an angle!

Figure 18.7

A circle with center O, chord $\overline{AB}$ with midpoint M and perpendicular bisector l, radius $\overline{OD}$ which passes through M, and radii $\overline{OA}$ and $\overline{OB}$.

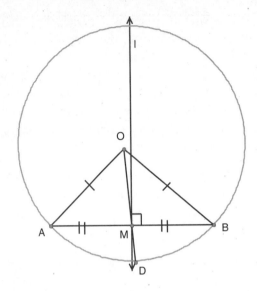

Because $\overline{OD}$ bisects our chord, it passes through M. You can use the SSS Postulate to show that $\triangle AOM \cong \triangle BOM$. That means that $\angle AMO \cong \angle BMO$. Well, $\angle AMO$ and $\angle BMO$ are also supplementary angles, because $\angle AMB$ is a straight angle. Using a little algebra, you can show that $\angle AMO$ and $\angle BMO$ are right angles. That means that $\overline{OM}$ is perpendicular to $\overline{AB}$. Because $\overline{OM}$ and your line are both perpendicular to $\overline{AB}$ and both pass through M, they must coincide. (Remember Theorem 10.1: Given a point A on a line l, there exists a unique line m which is perpendicular to l and passes through A). That contradicts the assumption that the perpendicular bisector of the chord is not a radius of the circle.

Let's write it up.

Example 4: Write a two-column proof of Theorem 18.7.

Solution: We'll start with what we are given.

Given: A circle with center O, a chord $\overline{AB}$ with midpoint M, and a perpendicular bisector line l.

Prove: $\overline{OM} \perp \overline{AB}$ and bisects $\overline{AB}$.

Proof: Here are the details.

Statements	Reasons
1. A circle with center O, a chord $\overline{AB}$ with midpoint M, and a line l which passes through M and is perpendicular to $\overline{AB}$.	Given

Statements	Reasons
2. $\overline{AM} \cong \overline{MB}$	Definition of midpoint
3. $\overline{OD}$ is a radius of the circle which passes through M	Construction
4. $\overline{OD}$ bisects $\overline{AB}$	Definition of segment bisector
5. $\overline{OA}$ and $\overline{OB}$ are also radii of our circle	Definition of radii
6. $\overline{OA} \cong \overline{OB}$	Theorem 17.1
7. $\overline{OM} \cong \overline{OM}$	Reflexive property of $\cong$
8. $\triangle AOM \cong \triangle BOM$	SSS Postulate
9. $\angle AMO \cong \angle BMO$	CPOCTAC
10. $m\angle AMO = m\angle BMO$	Definition of $\cong$
11. $\angle AMB$ is a straight angle, and $m\angle AMB = 180°$	Definition of straight angle
12. $m\angle AMO + m\angle BMO = m\angle AMB$	Angle Addition Postulate
13. $m\angle AMO + m\angle BMO = 180°$	Substitution (steps 11 and 12)
14. $2m\angle AMO = 180°$	Substitution (steps 10 and 13)
15. $m\angle AMO = 90°$	Algebra
16. $\angle AMO$ is right	Definition of right angle
17. $\overline{OM}$ is a segment passing through M and $\overline{OM} \perp \overline{AB}$	Definition of $\perp$
18. $\overline{OM}$ lies on l	Theorem 10.1

That's all of the proofs for now. Of course, inductive reasoning indicates that there will be more to prove in Chapter 19.

Putting the Pieces Together

As a grand finale, let's work through a couple of applications of this theorem, just to see how all of the pieces fit into place.

Example 5: Given a circle with center O and a circle with center P as shown in Figure 18.8, if $\overline{OP} \perp \overline{AB}$, M is the point of intersection of $\overline{OP}$ and $\overline{AB}$, and OA = 10, PA = 10, and AB = 12, find OP.

Figure 18.8

A circle with center O and a circle with center P with $\overline{OP} \perp \overline{AB}$, and M is the point of intersection of $\overline{OP}$ and $\overline{AB}$.

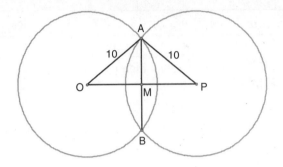

AB=12

Solution: Because $\overline{OP} \perp \overline{AB}$, we know that $\overline{OP}$ bisects $\overline{AB}$. This is because $\overline{AB}$ is a chord of the circle with center O, and $\overline{OP}$ is an extension of a radius of this circle. By Theorem 18.5, you know that $\overline{OP}$ bisects $\overline{AB}$, and AM = 6. Also, you have a right triangle, so you can use the Pythagorean Theorem to find OM:

$(OM)^2 + (AM)^2 = (OA)^2$

$(OM)^2 + 6^2 = 10^2$

$(OM)^2 = 64$

OM = 8

If you turn to the circle centered at P, you will have the same situation: $\triangle PMA$ is a right triangle, and the Pythagorean Theorem enables you to find PM:

$(PM)^2 + (AM)^2 = (PA)^2$

$(PM)^2 + 6^2 = 10^2$

$(PM)^2 = 64$

PM = 8

You are now at the point where you can answer the question. Using the Segment Addition Postulate (Postulate 3.5), OP = OM + MP = 8 + 8 = 16.

Example 6: This is the area of the pizza problem that I promised you. Suppose a pizza has an 8-inch radius, and you eat a slice (let's assume that this pizza was professionally sliced along diameters) with central angle having a degree measure of 60°. To help you visualize your lunch, I've drawn your pizza (you will have to imagine the sauce, cheese and toppings) in Figure 18.9. You eat the pizza enclosed in $\triangle ABO$ and throw away the crust. Find the area of the sector ABO, $\triangle ABO$, and the sliver of crust that you threw away.

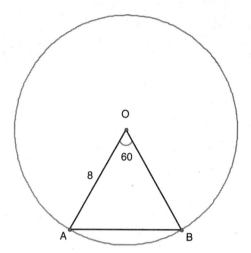

Figure 18.9

A circle with center O, *radii*
$\overline{OA}$ *and* $\overline{OB}$.

Solution: Find the area of the sector first: The area is given by
$$\frac{\theta \pi r^2}{360} = \frac{60\pi(8)^2}{360} = \frac{64\pi}{6} = \frac{32\pi}{3} \text{ inches}^2.$$

This is roughly 33.5 inches². Next tackle the area of the triangle. The area of a triangle is one-half the base times the height. Unfortunately, you don't know the base or the height of this triangle. But two of the three sides of our triangle are radii and, hence, are congruent. So you have an isosceles triangle. The angles opposite the congruent sides are congruent: $\angle A \cong \angle B$. You can determine the measures of the other two angles:

$$m\angle A + m\angle B + m\angle O = 180°$$

$$2m\angle A + 60° = 180°$$

$$2m\angle A = 120°$$

$$m\angle A = 60°$$

So all three angles measure 60°, and you have an equiangular, hence an equilateral, triangle. We've analyzed those before: bisecting the central angle will produce two 30-60-90 triangles, each with a hypotenuse 8 inches in length. You spent a good bit of time learning how to find the lengths of the legs of a 30-60-90 triangle if you know the length of the hypotenuse. Recall that the shorter leg has length equal to one-half the length of the hypotenuse, and the longer leg has length equal to the product of the length of the shorter leg and the square root of three. So the base of your triangle is 8 inches, and the height is $4\sqrt{3}$. Therefore the area of your triangle is

$$\frac{1}{2}(8)\left(4\sqrt{3}\right) = 16\sqrt{3} \text{ inches}^2.$$

This number is roughly 27.7 inches². Looking at Figure 18.9, to find the area of the sliver of crust, just subtract the area of the triangle that you ate from the area of the sector: 33.5 inches² – 27.7 inches² = 5.8 inches². Now you know how much of your pizza you threw away.

Put Me in, Coach!

Here's your chance to shine. Remember that I am with you in spirit and have provided the answers to these questions in Appendix A.

1. Write a formal proof of Theorem 18.2.

2. Write a formal proof of Theorem 18.4.

3. Write a formal proof of Theorem 18.6.

The Least You Need to Know

◆ Congruent chords have congruent arcs, and vice versa.

◆ A radius that is perpendicular to a chord bisects that chord.

◆ Combining your familiarity with triangles with your knowledge of triangles you can work out some impressive calculations.

Chapter 19

Circular Arguments

In This Chapter

- ◆ Relating chords and arcs
- ◆ Finding similar triangles in circles
- ◆ Properties of parallel chords
- ◆ Applying what you've learned

The shampoo industry certainly hopes you will read the instructions on their bottles. If you did, you would lather, rinse, and repeat. Then you would lather, rinse, and repeat. Then you would lather, rinse, and repeat .... You would use an entire bottle every time you washed your hair. I'm sure that the advertising firm that came up with that idea was sure they would cash in big. They didn't think about how infrequently people actually read instructions.

When you think of circular arguments, you probably imagine an argument that never ends. You are caught in an infinite loop, with no hope of escape. You can also think of them as arguments that go both ways. For example, if you want to learn geometry, you will read this book. And if you read this book, it's because you want to learn geometry.

You have seen examples of circular arguments throughout this book. For example, you know that when two parallel lines are cut by a transversal, the alternate interior angles are congruent. On the other hand, when two lines are cut by a transversal so that alternate interior angles are congruent, then the two lines are parallel. Think back to the various theorems that you have proven. How many of them are circular?

Angles and Arcs

It might not come as a great surprise that if two arcs are congruent, so are their central angles. It might also be obvious that if two central angles are congruent, so are their intercepted arcs. You can prove both of these statements with very little effort.

Theorem 19.1: If two arcs are congruent, their central angles are congruent.

Theorem 19.2: If two central angles are congruent, their intercepted arcs are congruent.

I'll take the first theorem, and you can prove the second one.

Example 1: Prove Theorem 19.1.

Solution: I'll need a circle with two congruent arcs, as shown in Figure 19.1.

Figure 19.1

A circle with $\overset{\frown}{AC} \cong \overset{\frown}{DE}$.

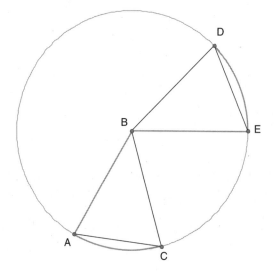

Given: $\overset{\frown}{AC} \cong \overset{\frown}{DE}$

Prove: $\angle ABC \cong \angle DBE$.

Proof: Make use of the Central Angle Postulate (Postulate 17.1): The degree measure of a central angle is equal to the degree measure of its intercepted arc. There is really not that much to prove here.

	Statements	Reasons
1.	A circle with center B and $\overset{\frown}{AC} \cong \overset{\frown}{DE}$	Given
2.	$m\overset{\frown}{AC} = m\overset{\frown}{DE}$	Definition of $\cong$
3.	$m\overset{\frown}{AC} = m\angle ABC$ and $m\overset{\frown}{DE} = m\angle DBE$	Central Angle Postulate
4.	$m\angle ABC = m\angle DBE$	Substitution (steps 2 and 3)
5.	$\angle ABC \cong \angle DBE$	Definition of $\cong$

That wasn't bad at all. The proof of your theorem shouldn't be much different.

Next, suppose you have two generic, run-of-the-mill chords with no special relationship whatsoever, other than that they intersect inside of the circle. They are not diameters or radii, congruent, parallel, perpendicular, or anything like that. They are just a couple of chords which happen to meet inside the circle. Let's call their point of intersection E.

Now, don't assume that E is the center of the circle or jump to any other conclusions about E. E is just a point inside the circle, and both chords happen to pass through it. Because the chords intersect inside the circle (Figure 19.2 illustrates what is going on here), they form some angles (four of them, actually). There are two pairs of vertical angles formed, and the fact that $\angle AEB \cong \angle CED$ is yesterday's news. But if you knew $m\overset{\frown}{AFB}$ and $m\overset{\frown}{CGD}$ and had to find me$m\angle AEB$, what would you do?

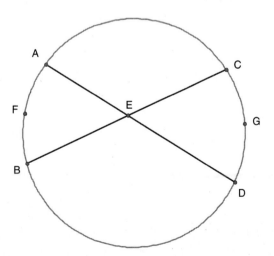

Figure 19.2

A circle with chords $\overline{AD}$ and $\overline{BC}$ that intersect at E.

After you state and prove this next theorem, you'll see how easy it would have been to solve the problem I posed.

Theorem 19.3: The measure of an angle formed by two chords that intersect within a circle is found by taking the average of the degree measures of the arcs intercepted by the angle and its vertical angle.

To translate this theorem into practical steps, if you know $m\overset{\frown}{AFB}$ and $m\overset{\frown}{CGD}$, all you have to do is average those degree measures and you know $m\angle AEB$. Let's prove it and then use it.

> **Example 2**: Write a formal proof of Theorem 19.3.
>
> **Solution**: You already have a drawing of this situation, thanks to Figure 19.2. Let's interpret this theorem in terms of this drawing.
>
> Given: A circle with chords $\overline{AD}$ and $\overline{BC}$ that intersect at a point E.
>
> Prove: $m\angle AEB = \frac{1}{2}(m\overset{\frown}{AFB} + m\overset{\frown}{CGD})$

Proof: You need a game plan. You know about the relationship between central angles and intercepted arcs and about the relationship between inscribed angles and intercepted arcs. What you don't know about are angles with vertices inside the circle that aren't at the center of the circle.

You need to relate the arcs and angles in the figure to arcs and angles that you understand. To do this, you'll need to change your perspective. Let's draw another chord: $\overline{BD}$, as shown in Figure 19.3. Now you have some inscribed angles to go with your intercepted arcs: $\angle ADB$ is an inscribed angle with intercepted arc $\overset{\frown}{AFB}$, and $\angle CBD$ is an inscribed angle with intercepted arc $\overset{\frown}{CGD}$.

Figure 19.3

A circle with chords $\overline{AD}$ and $\overline{BC}$ that intersect at E, and chord $\overline{BD}$.

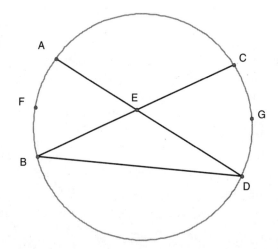

You know how the angles are related to the degree measures of their intercepted arcs: The measure of an inscribed angle is equal to one-half of the degree measure of its intercepted arc. Not only that, but $\angle$AEB is an exterior angle to $\triangle$BED, and you know how exterior angles are related to nonadjacent interior angles: the measure of an exterior angle of a triangle is equal to the sum of the measures of the nonadjacent interior angles. I think you're ready for the columns.

Statements	Reasons
1. A circle with chords $\overline{AD}$ and $\overline{BC}$ that intersect at a point E.	Given
2. Construct the chord $\overline{BD}$	Construction
3. $\angle$ADB and $\angle$CBD are inscribed angles	Definition of inscribed angle
4. $m\angle ADB = \frac{1}{2} m\overset{\frown}{AFB}$, $m\angle CBD = \frac{1}{2} m\overset{\frown}{CGD}$	Theorem 17.2
5. $\angle$AEB is an exterior angle to $\triangle$BED	Definition of exterior angle
6. $m\angle AEB = m\angle CBD + m\angle ADB$	Theorem 11.3
7. $m\angle AEB = \frac{1}{2} m\overset{\frown}{CGD} + \frac{1}{2} m\overset{\frown}{AFB}$	Substitution (steps 4 and 6)
8. $m\angle AEB = \frac{1}{2}(m\overset{\frown}{AFB} + m\overset{\frown}{CGD})$	Algebra

Let's work out an example to see how this theorem can be used.

Example 3: If, as in Figure 19.3, $m\overset{\frown}{AFB} = 54°$ and $m\overset{\frown}{CGD} = 36°$, find $m\angle AEB$.

Solution: You can use the previous theorem, because you just finished proving it:

$$m\angle AEB = \frac{1}{2}(m\overset{\frown}{AFB} + m\overset{\frown}{CGD}) = \frac{1}{2}(54° + 36°) = \frac{1}{2}(90°) = 45°$$

Similarity

Let's go back to the situation where you have two generic, run-of-the-mill chords with no special relationship whatsoever, other than that they intersect inside of the circle. I'll redraw the diagram in Figure 19.4. You observed (and proved) a relationship between a pair of vertical angles formed and the degree measures of their intercepted arcs. But there are more relationships that you can establish.

Figure 19.4

A circle with two chords that intersect at E.

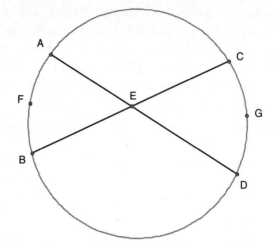

For example, you can create a couple of triangles, ΔABE and ΔCED. Now, I'm not saying that these are the only two triangles that you can make. They are just the two that I want to talk about right now. These two triangles have one pair of congruent angles, ∠AEB ≅ ∠CED. Wouldn't it be great if they had a second pair of congruent angles? Then the two triangles would be similar, and you could use CSSTAP.

Which couple of angles do you expect to be congruent? If you look at Figure 19.5, I've drawn the two triangles involved. Notice that ∠BAD and ∠BCD are both in-scribed angles that have the same intercepted arc. If any two angles are congruent, these are prime candidates! Now you've got two pairs of congruent angles, so you can apply the AA Similarity Theorem to conclude that the two triangles are similar. You can use CSSTAP to establish all kinds of proportionalities between the sides of these triangles. Let's write some theorems and share the burden of proof.

Figure 19.5

A circle with two chords that intersect at E, *forming two triangles,* ΔABE *and* ΔCED.

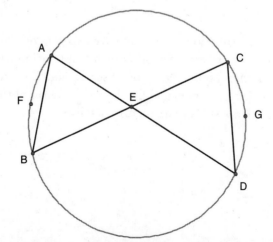

Theorem 19.4: If two chords intersect within a circle, then the product of the lengths of the segments of one chord is equal to the product of the lengths of the segments of the other.

That's some mighty fancy language for a theorem. Let's pick the terms apart and see what this theorem is trying to say. Going back to Figure 19.5, this theorem says that $AE \times ED = BE \times EC$ or, rearranging things,

$$\frac{AE}{EC} = \frac{BE}{ED}.$$

If you can show that $\triangle ABE \sim \triangle CDE$, then this proportionality will follow from CSSTAP.

Example 4: Prove Theorem 19.4.

Solution: You already have a drawing (Figure 19.5), so you just have to interpret this theorem in terms of the chords and angles in the diagram.

Given: A circle with chords $\overline{AD}$ and $\overline{BC}$ that intersect at E.

Prove: $AE \times ED = BE \times EC$.

	Statements	Reasons
1.	A circle with chords $\overline{AD}$ and $\overline{BC}$ that intersect at E	Given
2.	$\angle AEB$ and $\angle CED$ are vertical angles	Definition of vertical angles
3.	$\angle AEB \cong \angle CED$	Theorem 8.1
4.	$\angle BAD$ and $\angle BCD$ are inscribed angles with intercepted arc $\overset{\frown}{BFD}$	Definition of inscribed angle
5.	$m\angle BAD = \frac{1}{2}m\overset{\frown}{BFD}$ and $m\angle BCD = \frac{1}{2}m\overset{\frown}{BFD}$	Theorem 17.2
6.	$m\angle BAD = m\angle BCD$	Substitution (step 5)
7.	$\angle BAD \cong \angle BCD$	Definition of $\cong$
8.	$\triangle ABE \sim \triangle CDE$	Theorem 13.1: AA Similarity Theorem
9.	$\frac{AE}{EC} = \frac{BE}{ED}$	CSSTAP
10.	$AE \times ED = BE \times EC$	Algebra

You can use this theorem to solve for some chord lengths.

Example 5: If, in Figure 19.5, AE = 10, ED = 4, and EC = 8, find BE.

Solution: Use the proportionality developed in the proof of Theorem 19.4 to solve this problem:

$$\frac{AE}{EC} = \frac{BE}{ED}$$

$$\frac{10}{8} = \frac{BE}{4}$$

$$BE = 5$$

Let's analyze another situation involving intersecting chords. Suppose $\overline{AB}$ is a diameter of a circle and $\overline{CE}$ is a chord that is perpendicular to $\overline{AB}$. I've drawn you a picture in Figure 19.6. What can you say about the products of the lengths of the segments of the chords in this case?

Figure 19.6

A circle with diameter $\overline{AB}$ perpendicular to the chord $\overline{CE}$.

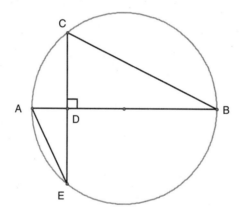

Because $\overline{CE}$ is perpendicular to $\overline{AB}$, you know that $\overline{AB}$ bisects $\overline{CE}$ and $\overline{CD} \cong \overline{DE}$. Because of Theorem 19.4, you know that

$$\triangle ADE \sim \triangle CDB, \text{ and } \frac{AD}{CD} = \frac{DE}{BD}$$

Combining these two facts, you have the proportion

$$\frac{AD}{CD} = \frac{CD}{BD}.$$

You've seen proportions before where the numerator of one side of the proportion is the same as the denominator of the other side of the proportion. In this case, CD is the geometric mean of AD and BD.

Parallel Chords and Arcs

When you had two chords that intersected *inside* of the circle, you were able to establish a relationship between the measures of the angles formed and the measures of the intercepted arcs. When two chords intersected *on* the circle, an inscribed angle was formed, and you proved that the measure of the inscribed angle is equal to one-half the degree measure of the intercepted arc. There is one more case that you need to consider. Suppose the two chords do not intersect at all. In other words, suppose the two chords are parallel. Figure 19.7 shows a circle with two parallel chords. These chords do not intersect, so they don't form an angle. But there are still some arcs that you can explore.

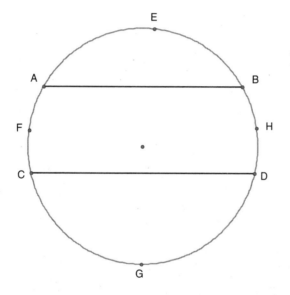

Figure 19.7

A circle with parallel chords $\overline{AB}$ *and* $\overline{CD}$.

The circle has been separated into four arcs: $\overset{\frown}{AEB}$, $\overset{\frown}{BHD}$, $\overset{\frown}{DGC}$ and $\overset{\frown}{CFA}$. You have to be careful to avoid being misled by the diagram. The chords can slide as close to the edge or as close to the center as they want. As the chords slide away from the center, the corresponding arcs get smaller and smaller. Because I can make these arcs as small as I want (of course, there is a limit to how big these arcs can get; I can't get any bigger than the entire circle!), there's no way that I can establish a definitive relationship between $\overset{\frown}{AEB}$ and $\overset{\frown}{DGC}$. But what about the other two arcs? Is there a relationship between $\overset{\frown}{BHD}$ and $\overset{\frown}{CFA}$? Maybe they are congruent? Let's explore these arcs in more detail. Somewhere along the way you should use the fact that the two chords are parallel. Hmmm … maybe one of the angle relationships involved with parallel lines cut by a transversal will come in handy.

You'll need a transversal. Let's construct the chord $\overline{BC}$. Now you have two parallel segments ($\overline{AB}$ and $\overline{CD}$) cut by a transversal ($\overline{BC}$), so you know that alternate interior angles are congruent: $\angle ABC \cong \angle BCD$. Well, those angles aren't just alternate interior angles; they are also inscribed angles! And the two arcs under investigation just happen to be intercepted arcs of those inscribed angles! How convenient. So, if the two inscribed (a.k.a. alternate interior) angles are congruent, so are the intercepted arcs. Now the skeptics are starting to come around. It looks like you have the logic to back up your hunch, and out of your hunch a theorem is born.

> **Theorem 19.5**: If two chords of a circle are parallel, then the intercepted arcs between these chords are congruent.

Example 6: Prove Theorem 19.5.

Solution: You already have a drawing, shown in Figure 19.7. You need to interpret what you are given and what you need to prove in terms of this drawing.

Given: A circle with chords $\overline{AB}$ and $\overline{CD}$, with $\overline{AB} \parallel \overline{CD}$.

Prove: $\overset{\frown}{AFC} \cong \overset{\frown}{BHD}$.

Proof: You already have a game plan. It's now a matter of putting this plan into columns.

	Statements	Reasons
1.	A circle with chords $\overline{AB}$ and $\overline{CD}$, with $\overline{AB} \parallel \overline{CD}$.	Given
2.	Construct the chord $\overline{BC}$	Construction
3.	$\overline{AB}$ and $\overline{CD}$ are parallel segments cut by transversal $\overline{BC}$	Definition of transversal
4.	$\angle ABC$ and $\angle BCD$ are alternate interior angles	Definition of alternate interior angles
5.	$\angle ABC \cong \angle BCD$	Theorem 10.2
6.	$m\angle ABC = m\angle BCD$	Definition of $\cong$
7.	$\angle ABC$ and $\angle BCD$ are inscribed angles, with intercepted arcs $\overset{\frown}{AFC}$ and $\overset{\frown}{BHD}$, respectively	Definition of inscribed angles
8.	$m\angle ABC = \frac{1}{2}m\overset{\frown}{AFC}$, and $m\angle BCD = \frac{1}{2}m\overset{\frown}{BHD}$	Theorem 17.2
9.	$\frac{1}{2}m\overset{\frown}{AFC} = \frac{1}{2}m\overset{\frown}{BHD}$	Substitution (steps 6 and 8)
10.	$m\overset{\frown}{AFC} = m\overset{\frown}{BHD}$	Algebra
11.	$\overset{\frown}{AFC} \cong \overset{\frown}{BHD}$	Definition of $\cong$

Putting Your Problems Behind You

Let's take a few moments to apply what you've learned. The more theorems you have learned, the more interesting are the problems you can solve. Start with a little algebraic exercise.

Example 7: If, in Figure 19.8, $m\angle AEC = 55°$, $m\overset{\frown}{AC} = 3x + 6$, and $m\overset{\frown}{BD} = x$, find mBD.

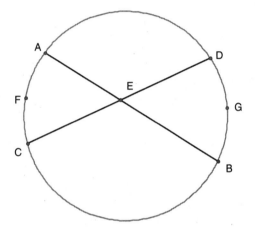

Figure 19.8

A circle with chords $\overline{AB}$ *and* $\overline{CD}$ *that intersect at* E.

Solution: The quick, easy, and incorrect answer is $m\overset{\frown}{BD} = x$. You need to know what number x is. To do that, you'll need to use some of your theorems from the last three chapters. The following result immediately comes to mind:

$$m\angle AEC = \tfrac{1}{2}\,(m\overset{\frown}{AFC} + m\overset{\frown}{DGB})$$

You can substitute into that equation:

$$55° = \frac{1}{2}(3x + 6 + x)$$

$$55° = \frac{1}{2}(4x + 6)$$

$$55° = 2x + 3$$

$$52° = 2x$$

$$x = 26°$$

Example 8: Suppose you have a circle with center O, as shown in Figure 19.9. If $\overline{OV}\perp\overline{RS}$, OV = 9 and OT = 6, find RS.

Figure 19.9

A circle with center O, and $\overline{OV}\perp\overline{RS}$.

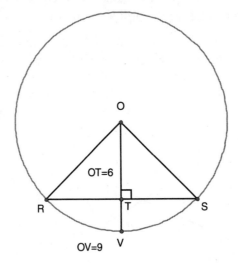

Solution: $\overline{OV}$. is a radius and $\overline{RS}$ is a chord. Because $\overline{OV}\perp\overline{RS}$, you know that $\overline{OV}$. bisects $\overline{RS}$. Find RT and double it to find RS. You also know that ΔORT is a right triangle and the hypotenuse is a radius of the circle. Because all radii are congruent, and you know the length of $\overline{OV}$.(another radius of the circle), you know that OR = 9. You know the length of the hypotenuse ($\overline{OR}$) and one of the legs ($\overline{OT}$), so you can use the Pythagorean theorem to find the length of the other leg ($\overline{RT}$). That's all there is to solving this problem:

$$(OT)^2 + (RT)^2 = (OR)^2$$

$$6^2 + (RT)^2 = 9^2$$

$$36 + (RT)^2 = 81$$

$$(RT)^2 = 45$$

$$RT = \sqrt{45} = 3\sqrt{5}$$

All you need to do is double the length of RT and you have RS:

$$RS = 2(RT) = 2\left(3\sqrt{5}\right) = 6\sqrt{5}$$

Put Me in, Coach!

Here's your chance to shine. Remember that I am with you in spirit and have provided the answers to these questions in Appendix A.

1. Write a formal proof of Theorem 19.2.

2. In Figure 19.10, suppose that $\overline{AB}$ and $\overline{CD}$ are diameters of a circle. If m∠AOC = 40°, find m$\overset{\frown}{DEB}$.

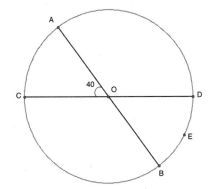

Figure 19.10

A circle with diameters $\overline{AB}$ and $\overline{CD}$.

3. If, as in Figure 19.3, m$\overset{\frown}{AFB}$ = 74° and m$\overset{\frown}{CGD}$ = 65°, find m∠AEB .

4. If, in Figure 19.5, AE = 15, AB = 12, CD = 4, and BE = 6, find $\overset{\frown}{EC}$ and $\overset{\frown}{ED}$.

5. If, as shown in Figure 19.11, you have a circle with center O, $\overline{OB}\perp\overline{AC}$, OA = 8, and OD = 6, find AC.

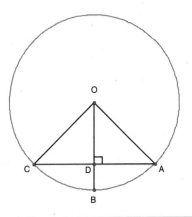

Figure 19.11

A circle with center O, $\overline{OB}\perp\overline{AC}$, OA = 8 and OD = 6.

The Least You Need to Know

- ◆ Congruent arcs have congruent central angles, and vice versa.

- ◆ To find the measure of an angle formed by two chords that intersect within a circle, take the average of the degree measures of the arcs intercepted by the angle and its vertical angle.

- ◆ If two chords intersect within a circle, then several angle relationships can be found.

- ◆ If two chords of a circle are parallel, then the intercepted arcs between these chords are congruent.

Chapter 20

The Unit Circle and Trigonometry

In This Chapter

- The tangent ratio
- The sine ratio
- The cosine ratio, and the rest
- How trigonometry relates to the unit circle

Trigonometry is the study of triangles. You might be wondering why I would write a chapter about trigonometry and put it in a section about circles. The immediate answer that comes to mind is that I just now remembered to write it and stuck it in wherever it would fit. Although I have been accused of being an absent-minded professor, when it comes to the writing of this book I have been on top of my game.

The reason that I am writing this chapter now is because so much about trigonometry has to do with circles! If I kept circles out of trigonometry, you wouldn't have the tools necessary to handle angles greater than 90°. You'll see what I mean once you embed these triangles inside the unit circle.

You will embark on your trigonometric travels via right triangles. Be sure to pack along the Pythagorean Theorem (our currency of choice) and properties of proportionalities (to convert currencies). You will then come full circle when you re-examine these ideas through the eyes of the unit circle.

The Tangent Ratio

Trigonometry got its start from calculations of early mathematicians and astronomers. They studied ratios of the lengths of the sides of right triangles and noticed some interesting things. For example, they observed that whenever the ratio of a shorter leg's length to the longer leg's length was 7/10 the angle opposite the shorter leg (that is, the smallest angle of the right triangle) was always 35°. Figure 20.1 shows a few triangles with leg length ratio 7/10. Notice that all of these triangles are similar. They each have a right angle, and they each have an angle of measure 35°. By the AA Similarity Theorem (Theorem 13.1), they are similar.

Figure 20.1

Three right triangles, each having the ratio of the length of the shorter leg divided by the length of the longer leg equal to ⁷/₁₀.

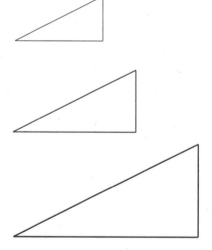

But this doesn't just happen with the ratio 7/10. All right triangles with the same leg length ratio will be similar to each other.

Figure 20.2 shows a typical right triangle. Recall that $\overline{AB}$ is the hypotenuse of the triangle. If you focus on $\angle BAC$, then $\overline{BC}$ is the side opposite $\angle BAC$ and $\overline{AC}$ is the side adjacent to $\angle BAC$. If your focus was on $\angle ABC$, then the opposite side is $\overline{AC}$ and the adjacent side is $\overline{BC}$. The nice thing about the hypotenuse is that it is always the side opposite the right angle, and that never changes. The ratio of the length of the opposite side of an angle divided by the length of the adjacent side in a right triangle is called the *tangent* of the angle. Write

$$\tan \angle BAC = \frac{BC}{AC}$$

when you want to say "the tangent of $\angle BAC$ is the ratio $\frac{BC}{AC}$." When there is no confusion about the angle involved, you can shorten this notation to

$$\tan \angle A = \frac{BC}{AC} \; .$$

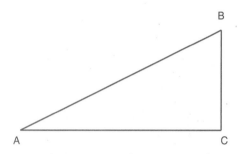

Figure 20.2

A right triangle $\triangle ABC$.

Notice that in the triangle in Figure 20.2, $\tan \angle A = \frac{BC}{AC}$ and $\tan \angle B = \frac{AC}{BC}$. These tangent ratios are the reciprocals of each other! That means that if $\tan \angle A > 1$, $\tan \angle B < 1$, and vice versa.

If the tangent ratio of an angle is equal to 1, that means that the length of the opposite side equals the length of the adjacent side. You have a triangle with two legs of equal length. You've seen that triangle before; it's a 45-45-90 triangle.

If you interpret this triangle as an object (say a tree) making a shadow on the earth (as shown in Figure 20.3), then this 35° angle is referred to as the angle of elevation.

Solid Facts

In a right triangle, the **tangent** of an angle is the ratio of the length of the opposite side of an angle divided by the length of the adjacent side.

You have seen an application of this idea earlier in the book. Remember the find-the-tree-height problem in Chapter 13? You solved this problem using similar triangles. You had the triangle formed by the tree and its shadow, and you had the triangle formed by you and your shadow. Using proportionalities, you were able to accurately estimate the height of the tree.

Figure 20.3

A tree making a shadow on the earth can be interpreted as forming a right triangle.

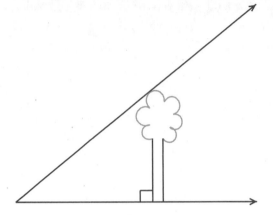

You can change the situation a little. What if you did not know your height, but could estimate the angle of elevation?

Example 1: Suppose that you have a tree, similar to the one shown in Figure 20.3. You are told that the angle of elevation is 30° and that $\tan 30° = .577$. If the shadow of the tree is 25 feet, estimate the height of the tree.

Solution: You know $\tan 30° = .577$. You also know that $\tan 30° = \dfrac{BC}{AC}$. Using substitution (and the fact that AC = 25), you have

$$\frac{BC}{25} = .577$$

$$BC = 25(.577) = 14.4 \text{ feet.}$$

Tangent Line
I have used the word tangent when referring to a line that just glances off of a circle. Now I'm talking about a tangent of an angle. Am I using the same word in two different contexts because I've run out of words and have to start using the same words for different ideas? Or is there a deeper connection that might become more apparent as the book progresses?

Early mathematicians calculated tables of values for the tangent of various functions. When I learned trigonometry (many years ago), my textbook came with tables of tangent values for various angles. Calculations were much more tedious, because we didn't have calculators (not even the basic kind that just add, subtract, multiply, and divide).

These days, trig books don't even mention the existence of these tables, much less include them. That's because any scientific calculator (the only ones worth spending money on) has a tangent button (TAN). Depending on the kind of calculator you own, you will either just enter your angle measure and push the TAN button, or you'll push the TAN button, the angle measure, and the "enter" or "=" button. You'll get to practice finding the tangent of a variety of angles as you go on.

The Sine Ratio

The tangent function examined the values of the ratio of the two legs of the right triangle. You can also take the ratio of the length of one leg divided by the length of the hypotenuse. You have two legs, so which one should you use?

In a right triangle, every angle has an opposite side and an adjacent side. And there's always a hypotenuse. Don't forget the hypotenuse. If you work with the opposite side and the hypotenuse you are dealing with the sine ratio. The *sine* of an angle is the ratio of length of the opposite side divided by the length of the hypotenuse.

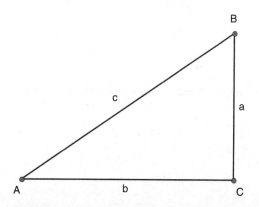

Solid Facts

In a right triangle, the **sine** of an angle is the ratio of the length of the opposite side divided by the length of the hypotenuse.

Let's use the Pythagorean Theorem to explore some of the properties of the sine ratio. Given a right triangle (like the one shown in Figure 20.4), let the lengths of the triangle be a, b, and c, where a is the length of the side opposite $\angle A$, b is the length of the side opposite $\angle B$, and c is the length of the side opposite $\angle C$ (and is the length of the hypotenuse). Then the Pythagorean Theorem tells you that $a^2 + b^2 = c^2$. The sine of $\angle A$ (which will be written $\sin \angle A$) is the ratio of the length of the side opposite $\angle A$ divided by the length of the hypotenuse; $\sin \angle A = \dfrac{a}{c}$. Because $a \leq c$, $\sin \angle A \leq 1$ (the only way $\sin \angle A = 1$ is if a = c, but that would make for a strange triangle!), the sine ratio cannot be greater than 1.

B

Figure 20.4

A right triangle with side lengths a *and* b, *and hypotenuse length* c.

c

a

A   b   C

If you are given a right triangle, and you know the tangent of the angle, you can find the sine of the angle by applying the Pythagorean Theorem.

> **Tangent Line**
>
> Let's explore the sine ratio for a right angle. The definition of the sine ratio is the ratio of the length of the opposite side divided by the length of the hypotenuse. Well, the length of the side opposite $\angle C$ is the length of the hypotenuse, so $\sin \angle C = \dfrac{c}{c} = 1$
>
> Because $\angle C$ is a right angle, $m\angle C = 90°$, so $\sin 90° = 1$.

Example 2: Suppose a right triangle has an angle with tangent ratio $\frac{5}{12}$. Find the sine ratio of that angle.

Solution: Figure 20.5 will help you visualize what is going on.

You are given that $\tan \angle A = \frac{5}{12}$, so a = 5 and b = 12. You can use the Pythagorean Theorem to find the length of the hypotenuse:

$c^2 = a^2 + b^2$

$c^2 = 5^2 + 12^2 = 25 + 144 = 169$

c = 13

Figure 20.5

A right triangle with $\tan \angle A = \frac{5}{12}$.

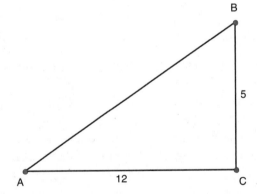

Now that you know the length of the hypotenuse, it is a simple matter of finding the sine ratio of that angle (it's the ratio of the length of the side opposite the angle divided by the length of the hypotenuse):

$\sin \angle A = \frac{5}{13}$

This calculation works both ways. If you are given the sine ratio of an angle of a triangle, you can find the tangent ratio of the angle.

Example 3: If a right triangle has an angle with a sine ratio of $\frac{4}{5}$, find the tangent ratio of that angle.

Solution: Let's take a look at Figure 20.6. In order to find the tangent ratio of the angle, you need to know the length of the side opposite the angle and the length of the side adjacent to the angle. Because the sine ratio is $\frac{3}{5}$, the length of the side opposite the angle is 3 and the length of the hypotenuse is 5. Using the Pythagorean Theorem, you can solve for the length of the side adjacent to the angle:

$a^2 + b^2 = c^2$

$3^2 + b^2 = 5^2$

$9 + b^2 = 25$

$b^2 = 16$

$b = 4$

So the tangent ratio is $\frac{3}{4}$.

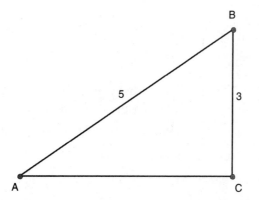

Figure 20.6

A right triangle with an angle having a sine ratio of $\frac{3}{5}$.

The Pythagorean Theorem will be used so often throughout this chapter that you will know it like the back of your hand by the time you are finished with this chapter!

The Cosine Ratio

The story so far: Given one of the acute angles in a right triangle, you have studied two ratios involving the lengths of the sides of the triangle. The tangent ratio involved the length of the side opposite the angle divided by the length of the side adjacent to the angle. The sine ratio involved the length of the side opposite the angle divided by the length of the hypotenuse of the triangle. You have been playing favorites with the opposite side of the angle at the expense of the adjacent side. To even things up, let me introduce a new ratio, the cosine ratio. The *cosine* of an angle is the ratio of the length of the side adjacent to the angle divided by the length of the hypotenuse of the triangle. The cosine of ∠A will be abbreviated as cos∠A .

You can play the same kinds of games that you played with the tangent and sine ratios.

Example 4: If a right triangle has an angle with tangent ratio $\frac{9}{14}$, find the sine ratio and the cosine ratio of the angle.

Solution: Because a picture is worth a thousand words when working out these problems, I have sketched this situation in Figure 20.7. Because you need to know the sine and cosine ratios of the angle, you will need to calculate the length of the hypotenuse of the triangle. You can whip out the Pythagorean Theorem and take care of that right now:

$$a^2 + b^2 = c^2$$

$$9^2 + 14^2 = c^2$$

$$c^2 = 277$$

$$c = \sqrt{277}$$

Solid Facts

In a right triangle, the **cosine** of an angle is the ratio of the length of the side adjacent to the angle divided by the length of the hypotenuse of the triangle.

Now that you know the length of all three sides, finding the sine and cosine ratios can be done using the definition $\sin \angle A = \frac{9}{\sqrt{277}}$ and $\cos \angle A = \frac{14}{\sqrt{277}}$.

Figure 20.7

A right triangle with tangent ratio $\frac{9}{14}$.

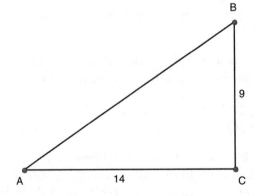

At this point, you might be tempted to rationalize the denominator and write your answers as

$$\sin \angle A = \frac{9\sqrt{277}}{277} \text{ and } \cos \angle A = \frac{14\sqrt{277}}{277} .$$

If so, then I would be impressed with your willingness to take things one step further, fearlessly going deeper into the torrential algebraic waters to write your answer in the form that I'm sure your algebra teacher emphasized when you first learned about radicals.

Of course, it is possible that the thought of rationalizing the denominator (as the process is officially called) never even occurred to you. That's okay, too. This isn't an algebra book, and there are advantages to leaving things in such a technically improper form. (Though when I wear my algebraic hat you'll never hear—or read—me say—or write—that it's okay to leave things improperly.)

One advantage to leaving things improperly is that you can clearly see that the sine and cosine ratios are less than 1. That's a quick and easy check to see if your answers make sense. The sine and cosine ratios can be equal to 1 on special occasions, but the ratios will never be greater than 1. Remember that the tangent ratio has no such restriction. You already saw that the tangent ratio can be greater, less than, or equal to 1.

You can work out these calculations in a variety of directions. If you are given the sine, cosine, or tangent ratio of an angle, you can find the other two ratios after using the Pythagorean Theorem.

> **Tangled Knot**
>
> The sine and cosine ratios of an angle cannot be greater than 1. The tangent ratio has no such restriction.

And the Rest

There are three other trigonometric ratios that I will just mention. These three new ratios actually come from the first three ratios.

The tangent of an angle is the ratio of the length of the side opposite the angle divided by the length of the adjacent side. The cotangent of the angle is just the reciprocal of the tangent ratio. The *cotangent* of an angle is the ratio of the length of the adjacent side divided by the length of the side opposite the angle.

The sine of an angle is the ratio of the length of the side opposite the angle divided by the length of the hypotenuse of the triangle. The cosecant of the angle is the reciprocal of the sine of the angle. The *cosecant* of an angle is the ratio of the length of the hypotenuse of the triangle divided by the length of the side opposite the angle.

The cosine of an angle is the ratio of the length of the adjacent side divided by the length of the hypotenuse of the triangle. The secant of the angle is the reciprocal of the cosine of the angle. The *secant* of an angle is the ratio of the length of the hypotenuse of the triangle divided by the length of the adjacent side.

Solid Facts _____

In a right triangle, the **cotangent** of an angle is the ratio of the length of the adjacent side divided by the length of the side opposite the angle.

In a right triangle, the **cosecant** of an angle is the ratio of the length of the hypotenuse of the triangle divided by the length of the side opposite the angle.

In a right triangle, the **secant** of an angle is the ratio of the length of the hypotenuse of the triangle divided by the length of the adjacent side.

Because the cotangent, cosecant, and secant ratios of an angle are just the reciprocals of the tangent, sine, and cosine ratios, the properties of these new ratios fall out from the properties of the old ratios. For example, because the sine of an angle cannot be greater than 1, the cosecant of an angle cannot be less than 1.

There is an important relationship between the sine and cosine ratios that comes from the Pythagorean Theorem. Given a right triangle, like that shown in Figure 20.8, the Pythagorean Theorem tells us $a^2 + b^2 = c^2$. If you divide both sides of this equation by c^2, you have

$$\left(\frac{a}{c}\right)^2 + \left(\frac{b}{c}\right)^2 = 1.$$

Because

$$\sin \angle A = \frac{a}{c} \text{ and } \cos \angle A = \frac{b}{c},$$

Eureka! _____

Given any angle in a right triangle, $(\sin\angle A)^2 + (\cos\angle A)^2 = 1$.

you can substitute these trigonometric ratios into the equation: $(\sin \angle A)^2 + (\cos \angle A)^2 = 1$. It turns out that this relationship between the sine and cosine ratios hold for any angle! This relationship is used repeatedly in trigonometry, and it becomes as important as the Pythagorean Theorem. Because it essentially comes from the Pythagorean Theorem, that shouldn't surprise you.

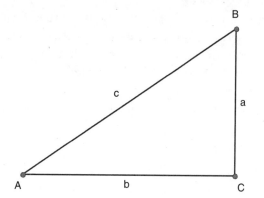

Figure 20.8

A right triangle △ABC.

How Does This Relate to the Unit Circle?

Our trigonometric ratios were defined within the confines of a right triangle. Because the measures of the interior angles of a right triangle add up to 180°, and one of these angles has a measure equal to 90°, the other two angles of a right triangle must be acute angles. So you can only find the trigonometric ratios of acute angles. That's too limiting.

The sine, cosine, and tangent ratios can be defined for any angle, not just acute angles. But in order to do this, you must embed a right triangle into a circle. Although you could use any circle, things work out nicely if you use the "unit circle." You might be wondering what the unit circle is and why you should use that particular circle. Well, the unit circle is a circle with radius equal to 1. Let's embed a right triangle into a unit circle and see what happens. Use Figure 20.9 as a guide.

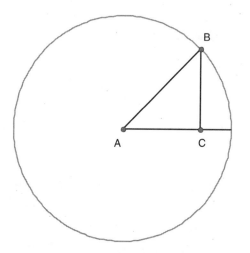

Figure 20.9

A right triangle embedded in the unit circle.

Each one of the vertices of the triangle will have some special characteristic, but the coveted properties are spread around so that one vertex isn't seen as being "better" than any other. Keep vertex A at the center of the circle. It is not allowed to move from that point. Think of vertex A as being in a trigonometric "time out." Vertex B is constrained to lie on the circle. It can move around the circle and occupy whatever place it wants, but it must stay on the circle. The only nice property left is the triangle's right angle, so by process of elimination ∠C is the right angle. There is one last restriction. Keep the diameter that C lies on fixed (in this case, $\overline{XY}$), and let B move around the circle.

The hypotenuse of your triangle has one end point located at the center of the circle and the other end point on the circle. Because you are working with the unit circle, your hypotenuse has length 1. Recall that the sine and cosine ratios both involve the length of the hypotenuse in the denominator. You can't have a nicer denominator of a ratio than 1. That's the advantage of working on the unit circle.

Now, let B move around the circle and always embed a right triangle by dropping a perpendicular line segment from B to $\overline{XY}$. In Figure 20.10, I've shown four different locations for B and the corresponding embedded right triangles that result. Notice that the orientation of the right triangle changes, but the hypotenuse is always the line segment $\overline{AB}$.

Figure 20.10

Four right triangles embedded in the unit circle, based on the location of the vertex B.

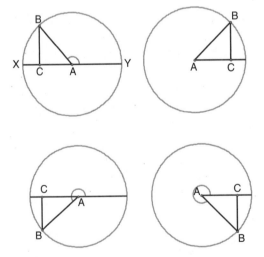

Whenever you talk about ∠A, you should always refer to ∠BAC, an interior angle of your triangle. The trigonometric ratios will always be

$$\tan \angle A = \frac{a}{b}, \sin \angle A = \frac{a}{c}, \text{ and } \cos \angle A = \frac{b}{c}.$$

These will always be positive numbers, because the length of any side is positive (by the Ruler Postulate). These ratios can be simplified a bit because you are constrained to working on the unit circle:

$$\tan \angle A = \frac{a}{b} \text{ , } \sin \angle A = a \text{, and } \cos \angle A = b$$

Now that the stage is set, I am ready to talk about the trigonometric ratios of any angle (not just acute angles)! I'll divide the circle into quarters and discuss angles that fall into each quarter one at a time. The angles that fall in the first quarter are acute angles, and I have already discussed the trigonometric ratios of acute angles.

For the second quarter, suppose you have an obtuse angle θ, as shown in Figure 20.11. Then the supplement of this obtuse angle is an acute angle. Embed a triangle like that in Figure 20.10(b) and define the trigonometric ratios of θ based on the trigonometric ratios of $\angle A$: $\sin \theta = \sin \angle A$, $\cos \theta = -\cos A$, and $\tan \theta = -\tan \angle A$. Notice that the sine ratio of an obtuse angle has the same numerical value as the sine ratio of its acute supplement, whereas the cosine and tangent ratios of an obtuse angle have the same absolute value as the cosine and tangent ratios of its supplement, but they have the opposite sign.

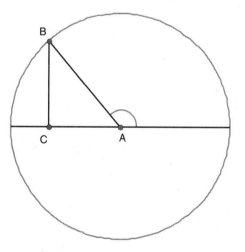

Figure 20.11

A circle with an obtuse central angle, and the corresponding embedded triangle.

Let's keep going around the circle. The third quarter involves angles whose measure is between 180° and 270°, as shown in Figure 20.12. Embed a triangle in this circle similar to the one shown in Figure 20.10(c), and again define the trigonometric ratios of these type of angles based on the trigonometric ratios of acute angles. All you need to do is change a few signs (two signs, actually): $\sin \theta = -\sin \angle A$, $\cos \theta = -\cos A$, and $\tan \theta = \tan \angle A$. Notice that the signs of the sine and cosine ratio are negative, and the tangent ratio is the positive one.

Figure 20.12

A circle with central angle between 180° and 270°, and the corresponding embedded triangle.

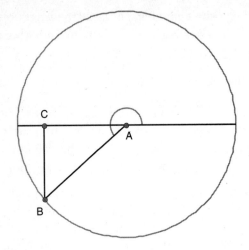

One more quarter and you graduate! This last quarter involves angles whose measure is between 270° and 360°, as shown in Figure 20.13. In this case, embed a right triangle like you did in Figure 20.10(d), and define the trigonometric ratios of the angle based on the trigonometric ratio of its corresponding acute interior angle. Once again, two signs will be changed: $\sin \theta = -\sin \angle A$, $\cos \theta = \cos A$, and $\tan \theta = -\tan \angle A$. This time the cosine ratio gets to stay positive, and the sine and tangent ratios are negative.

Figure 20.13

A circle with central angle between 270° and 360°, and the corresponding embedded triangle.

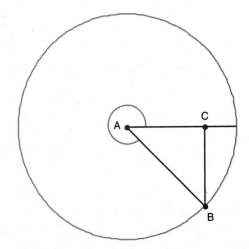

Eureka!

The trigonometric ratios of nonacute angles have the same magnitude as the trigonometric ratios of their corresponding acute angles. Two of the three ratios are negative, and one stays positive. As you go around the circle counter-clockwise, start with ALL ratios being positive, then the SINE ratio stays positive, next the TANGENT ratio is positive, and finally the COSINE ratio gets its turn. If you have trouble remembering which ratio is positive where, there is a famous mnemonic available to help: **A**ll **S**tudents **T**ake **C**alculus—**A**ll **S**ine **T**angent **C**osine.

This is just the tip of the trigonometric iceberg. I could fill an entire book with just the material discussed in a trigonometry class! I'll get started on that book right after I finish this one!

Put Me in, Coach!

Here's your chance to shine. Remember that I am with you in spirit and have provided the answers to these questions in Appendix A.

1. If a right triangle has an angle with tangent ratio $\frac{3}{5}$, find the sine ratio of that angle.

2. If a right triangle has an angle with a sine ratio of $\frac{1}{2}$, find the tangent ratio of that angle.

3. If a right triangle has an angle with cosine ratio $\frac{3}{7}$, find the sine and tangent ratios of the angle.

4. If a right triangle has an angle with sine ratio $\frac{5}{9}$, find the tangent and cosine ratios of the angle.

The Least You Need to Know

- The tangent ratio of an acute angle is the ratio of the length of the opposite side divided by the length of the adjacent side.

- The sine ratio of an acute angle is the ratio of the length of the opposite side divided by the length of the hypotenuse.

- The cosine ratio of an acute angle is the ratio of the length of the adjacent side divided by the length of the hypotenuse.

- To determine the sine, cosine and tangent ratios of a nonacute angle, embed a right triangle in the unit circle. Depending on which quadrant of the circle the angle is in, two of the trigonometric ratios will be negative.

Part 5

Where Can We Go from Here?

You've been stuck in this plane for way too long. It's time to land and refuel before you start the next leg of your journey. It's a good thing there is a third dimension for you to stretch in. While you are stretching be sure to take in the local sights: prisms, pyramids, and the Platonic solids.

Take a minute to learn how to make the geometric shapes you've heard so much about. After that, algebra and geometry will fuse together as you study the Cartesian Coordinate System. You'll learn your way around some non-Euclidean geometries before you get back on the plane and move your constructed geometric figures around. After that, you can sit back and reflect on what you've learned. What a long, strange trip it's been.

The Next Dimension: Surfaces and Solids

In This Chapter

◆ Prisms

◆ Modernize the ancient pyramids

◆ Cylinders without the horsepower

◆ Fall in love with the Platonic solids

You've spent the first 20 chapters of this book working in two dimensions. Although you might be able to name a few two-dimensional objects, most of the objects you encounter in your day-to-day lives are three-dimensional. This book, and even a page in this book, has three dimensions: a length, a height, and a thickness.

Because geometry is useful in dealing with the objects you encounter in your day-to-day lives, it is necessary to spend some time talking about three-dimensional objects. Some of the most common three-dimensional objects are prisms, cylinders, cones, spheres, and pyramids. These are mathematical terms for sheets of paper, soda cans, dunce caps, marbles, and pyramids.

The basis of three-dimensional objects are two-dimensional shapes. You will take polygons and circles in one plane and connect them to other polygons, circles, or points that lie in a different (but parallel) plane. I like to think of this process as mathematical wing walking.

Prisms

Suppose you have a polygon, like maybe a triangle or a rectangle, and you clone it. Now you have two congruent polygons lying side by side. If your polygons don't get along, and the plane isn't big enough for the both of them, one of them will have to move. Go ahead and move one of your polygons into a new plane. Don't twist it, or rotate it, or turn it in any way. Just lift it (or lower it, if you'd prefer) to a new plane that is parallel to the old plane. It doesn't have to be directly over (or under, if you chose to lower it) the other polygon; you can move it sideways if the spirit moves you. The goal is to have the two polygons in parallel planes, with the same orientation.

It's a well-known fact that absence makes the heart grow fonder. Now that your polygons have been separated, they might miss each other. That's not to say that they want to live in the same plane again. They might just want to establish some sort of connection. After all, these are congruent polygons, and the corresponding vertices might want to maintain their connection. Let's draw line segments connecting each pair of corresponding vertices. The result is a "solid" figure known as a *prism*. Two prisms (one based on a triangle, the other based on a rectangle) are shown in Figure 21.1. Notice that the two polygons lie in parallel planes.

Figure 21.1

Two examples of prisms.

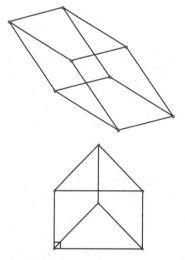

In order to talk about this new shape, I need to introduce you to some terms for the parts of a prism. The congruent polygons that lie in the parallel planes are the *bases* of the prism, and sides of these polygons are called the *base edges*. The line segments that connect the corresponding vertices of the congruent polygons are called the *lateral edges* of the prism. The polygons that are formed by the lateral edges are called the *lateral faces* of the prism. The vertices of the congruent polygons are the *vertices* of the prism. A prism will always have twice as many vertices as the base polygon.

Solid Facts

A **prism** is the region formed by two parallel congruent polygons having corresponding vertices joined by line segments.

The **bases** of a prism are the congruent polygons that lie in the parallel planes.

The **base edges** of a prism are the edges of the base of the prism.

A **vertex** of a prism is a point of intersection of two base edges.

The **lateral edges** of a prism are the line segments connecting corresponding vertices of the bases of the prism.

A **lateral face** of a prism is the quadrilateral formed by a pair of corresponding base edges and the lateral edges that connect the corresponding vertices.

In the triangular prism, one of the two congruent triangles is directly over the other one. In this case the lateral edges are perpendicular to the base edges, and the prism is called a *right prism*. If the bases of a right prism are regular polygons (recall that a regular polygon is an equilateral and equiangular polygon), then the prism is called a *regular prism*.

In the rectangular prism (the prism created from congruent rectangles) shown in Figure 21.1, the lateral edges are not perpendicular to the base edges. This is an example of an *oblique prism*.

By now you might have noticed that a prism's name is related to the shape of its base. A triangular prism has a triangular base; a rectangular prism has a rectangular base, and so on. This pattern will be followed to the end of time (or at least until your ability to name polygons expires).

Solid Facts

A **right prism** is a prism in which the lateral edges are perpendicular to the base edges at the vertices.

An **oblique prism** is a prism in which the angle formed between a lateral edge and a base edge is not a right angle.

A **regular prism** is a right prism whose base is a regular polygon.

The most familiar prism is a cube. A cube is a prism with a square base. But that's just the beginning! The lateral edges of a cube are perpendicular to the base edges, so a cube is a right prism. But wait! There's more! The bases of a cube are squares, which are regular polygons. So a cube is a right prism whose base is a regular polygon. In other words, a cube is a regular prism. But there's more to a cube than that. A cube has lateral edges that are the same length as the base edges. That's enough to put a cube into a very elite category. It's in such an elite class that there are only five shapes worthy of admission. You'll learn more about that class at the end of this chapter.

Eureka!

To name a prism, start with the name of the polygon that forms its base. Next, determine if the base of the prism is a regular polygon. If it is, you have a regular prism. If not, you just have a run-of-the-mill, generic prism. Finally, determine if the lateral edges are perpendicular to the base edges. If they are, you have a right prism; otherwise you have an oblique prism. Put all of the pieces together and you have named that prism.

Prisms are named and classified by their bases and the relationship between their lateral edges and their base edges. Let's practice.

Example 1: Name and classify each type of prism shown in Figure 21.2.

Figure 21.2

Three prisms to name and classify.

Solution: The first prism has a hexagonal base (not equilateral) and lateral edges perpendicular to the base edges. That makes it a right hexagonal prism. The second prism has a pentagonal base (not equilateral) and the lateral edges are not perpendicular to the base, so it's an oblique pentagonal prism. The third prism has a triangular base, and it is equilateral. By Theorem 14.3 it is also equiangular, which makes it regular. Because the lateral edges are perpendicular to the base edges, it is a right prism. Thus it is a right, regular, triangular prism.

Pyramids

A geometry book would not be complete if it did not include more than a passing reference to pyramids. The Egyptians have made pyramids famous, and people have traveled many miles to see these impressive structures. Engineers and archeologists are equally impressed with the design and construction of the pyramids.

You probably have an idea of what a typical *pyramid* looks like. It has a square base and four triangular sides rising up to join at a single point. It might come as a surprise to learn that from a mathematician's perspective, there are many kinds of pyramids. To construct a general pyramid, start with a polygon (it doesn't have to be a square, or even four-sided!) in a plane and a point above the polygon. Connect each vertex of the polygon to this point and voilà! You have a pyramid. The polygon that you started with is called the *base* of the pyramid, and the point is called the *apex* of the pyramid. I have drawn a traditional pyramid and a pentagonal pyramid for you in Figure 21.3. Notice that the lateral faces of a pyramid are triangles.

Solid Facts

A **pyramid** is the region formed by joining the vertices of a polygon within a plane to a point outside of the plane.

The **base** of a pyramid is the planar polygon whose vertices are all joined to the noncoplanar point.

The **apex** of the pyramid is the noncoplanar point.

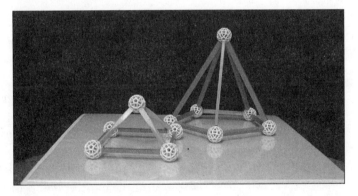

Figure 21.3

A square pyramid and a pentagonal pyramid.

Just as it is with prisms, the name of a pyramid is determined by the shape of its base. There are hexagonal pyramids and triangular pyramids. Try to reason out the polygon that forms the base of each of these pyramids. There are also regular pyramids: pyramids whose base is a regular polygon and whose lateral edges are all congruent.

Cylinders and Cones

In the definition of a prism, the bases were forced to be congruent polygons. If the bases are congruent closed curves (like a circle, for example) then you have a *cylinder*. I've sketched a cylinder with a circular base in Figure 21.4. The base of this cylinder is a circle so it doesn't have any vertices. Most of the cylinders you encounter have circles for the shape of the base, and they are called *circular cylinders*. But having a circular base is just a courtesy. It is not a requirement.

Figure 21.4

A circular cylinder.

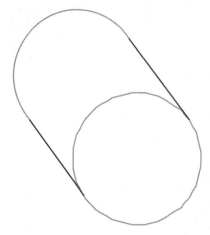

When you think of a cylinder, the immediate shape that comes to mind is that of a tin can. These types of cylinders are classified as right circular cylinders.

> **Solid Facts**
>
> A **cylinder** is the solid generated by two congruent closed curves in parallel planes together with the surface formed by line segments joining corresponding points of the two curves. A **circular cylinder** is a cylinder with a circular base.

Loosening up the restrictions on the bases of a prism created cylinders. Loosening up the restrictions on the shape of the base of a pyramid will result in the creation of a cone. Construct a cone the same way you constructed a pyramid. Start with a closed curve lying in a plane, which will serve as the *base* of the cone, and a point, which will serve as the *apex* of the cone. A *cone* is the solid formed by the interior of this closed curve together with the surface formed when each point on the closed curve is joined to the apex using line segments.

A cone is named based on the shape of its base. Figure 21.5 shows a circular cone. Circular cones fall into one of two categories: right circular cones and oblique circular cones. A *right circular cone* is a circular cone where the line segment connecting the apex of the cone to the center of the circular base is perpendicular to the plane of the base. *An oblique circular cone* is a circular cone where the line segment connecting the apex of the cone to the center of the circular base is not perpendicular to the plane of the base.

Figure 21.5

A circular cone.

Solid Facts

A **cone** is the solid formed by the interior of a planar closed curve together with the surface formed when each point on the closed curve is joined to a noncoplanar point using line segments.

The **base** of a cone is the planar closed curve of the cone.

The **apex** of a cone is the noncoplanar point that is joined to every point on the base.

A **right circular cone** is a circular cone where the line segment connecting the apex of the cone to the center of the circular base is perpendicular to the plane of the base.

An **oblique circular cone** is a circular cone where the line segment connecting the apex of the cone to the center of the circular base is not perpendicular to the plane of the base.

If you are having trouble visualizing the mathematical description of a cone, you can find a model of one at your favorite ice cream shop. Ask for your favorite flavor in the traditional serving container, a "sugar" cone. This is just a model of a cone, and some slight differences between the model and the real thing are bound to exist. For example, a mathematician's cone is solid, while the ice-cream cone is hollow. But you can avoid this difference by filling the cone with ice cream.

Polyhedra

A polyhedron is a three-dimensional version of a polygon. Recall that a polygon is the two-dimensional region that is bounded by line segments. When you move up in dimension, planes play the role of line segments. A polyhedron is a solid bounded by planar polygonal regions. The polygons that bound the polyhedron are called the faces of the polyhedron, and the edges of the polygon form the edges of the polyhedron. The vertices of the polygons are the vertices of the polyhedron. The prisms and pyramids discussed earlier are special types of polyhedra.

The prisms and pyramids discussed earlier are special types of polyhedra. Prisms have quadrilateral lateral surfaces and pyramids have triangular lateral faces. In general, a polyhedron can have any kind of polygon for any of its faces. They can be as wild as you can imagine them. But it turns out that your imagination is not as wild as you might think it is.

Leonhard Euler, a famous mathematician, discovered a special relationship between the number of vertices, edges, and faces that every polyhedron satisfies. That includes the biggest and the wildest polyhedron that the biggest and the wildest person can imagine. If you let V represent the number of vertices of a polyhedron, E stand for the number of edges of the polyhedron, and F denote the number of faces of the polyhedron, then $V + F = E + 2$. Let's take a simple polyhedron and see if it works.

Example 2: Verify Euler's formula for the cube shown in Figure 21.6.

Figure 21.6

A cube.

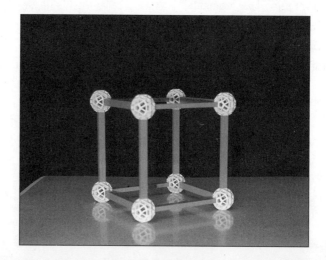

Solution: A cube has 8 vertices, 6 faces, and 12 edges. Let's see if it works: 8 + 6 = 12 + 2. Yep, it works!

Spheres

Thanks to early explorers, people have lost the fear of falling off the edge of the earth. It took quite awhile for people to accept the fact that the earth is round, but not circular. I mean that the earth is not the two-dimensional kind of round that CDs and plates are made of, but the three-dimensional kind of round that basketballs and oranges are made of. Even though the sun and the moon look like two-dimensional round objects up in the sky, you know that they have three dimensions of roundness. They are spherical: the three-dimensional equivalent of two-dimensional circles.

Tangent Line

Spheres are appealing for a variety of reasons. You can rotate a sphere or look at it from any vantage point and it will look the same. A sphere is one of the most symmetric shapes in nature. It is interesting that very few sports involve a perfectly symmetrical ball (Ping-Pong is one that immediately comes to mind). Asymmetries are introduced to change the features of the ball. For example, a golf ball has dimples and a baseball has stitches. But the dimples of a golf ball and the stitches of a baseball are strategically placed. New designs promise that your ball will travel farther and faster, and curve more or less (bending to your will, of course).

As with most three-dimensional objects, spheres are very similar to their two-dimensional counterparts. A circle is defined as the set of points in a plane that are a fixed distance r from a certain point, which is called the center of the circle. A sphere is defined in a similar manner, without the restriction that the points have to lie in one plane. A *sphere* is the set of all points (in three-dimensional space) that are a fixed distance r from a certain point, which is called the *center* of the sphere. The fixed distance r is the *radius* of the sphere.

Solid Facts

A **sphere** is the set of all points (in three-dimensional space) that are a fixed distance r from a certain point.

The fixed distance is called the **radius** of the sphere.

The **center** of the sphere is the point that is a fixed distance from each point on the sphere.

Spheres have diameters just like circles do. A diameter is a line segment connecting two points on the sphere that passes through the center. Circles can be cut by chords, so when this idea is generalized in three-dimensional space, planes will slice spheres. The intersection of a sphere and a plane is a circle.

Platonic Solids

Although a sphere has a symmetry that cannot be beat, there are some fairly symmetric flat-sided figures, as the polyhedra are often referred to. Remember the cube? It's a prism with a square base, with one base directly over the other one. It is formally known as a right square prism with a twist. All of the edges are congruent. If you turn a cube around and look at it from the same vantage point (a view along a particular edge, for example), all you see are squares. It's bursting at the edges with symmetry, and it belongs to an elite group of polyhedra: the Platonic Solids. The Platonic solids are the most symmetric group of solids around. There are only five of them, and there is no hope of inventing a sixth. Five is all there are, and five is all there'll be, 'til the end of time.

It's time to meet the other four Platonic solids. You'll know a Platonic solid by its symmetry; they are the most symmetric polyhedra around. Let's see what you can learn by looking at the cube. It is a right prism with a square base. Its base is a regular polygon: a polygon that is equilateral and equiangular.

There are lots of regular polygons hanging around, ready to be studied. You should start simple and work your way up. A cube started from a square, but that's not the simplest regular polygon. An equilateral triangle is the simplest regular polygon. Is it possible to build an incredibly symmetric polyhedron using only an equilateral triangle? As it turns out, three of the five Platonic solids are made using equilateral triangles.

Tangent Line
The cube is the only Platonic solid with a slang name. The cube is formally known as a hexahedron, because it has six faces.

The second Platonic solid (and the first using triangles) is made by constructing a triangular prism using congruent equilateral triangles for the base and the lateral faces. Its official name is a tetrahedron, because it has four faces. I've made one for you in Figure 21.7.

Figure 21.7

A tetrahedron.

There's more that you can do with triangles. The first triangular Platonic solid was a prism. The second triangular Platonic solid will come from a pyramid. Take four equilateral triangles and make the lateral surfaces of a square pyramid. Do this again, with four other equilateral triangles. Put your two structures together, as shown in Figure 21.8, and you have yet another triangular super-symmetric polyhedron. This polyhedron has eight faces and is called an octahedron.

Figure 21.8

An octahedron.

To make the third triangular Platonic solid, you will need more triangles. Twenty of them, to be precise. If you put twenty equilateral triangles together (you'll need lots of patience and tape as well), you will have created the fourth Platonic solid, the icosahedron.

You can take a break from working with triangles. You've already used the square, so it's time to take the next regular polygon on the list, a regular pentagon. Attach twelve pentagons together carefully, and you will have created a dodecahedron.

You can make a set of these Platonic solids using four congruent equilateral triangles for the tetrahedron, eight congruent equilateral triangles for the octahedron, and twenty congruent equilateral triangles for the icosahedron. To make the cube you will need six congruent squares, and to make the dodecahedron you will need twelve congruent pentagons. Create the Platonic solids, hold them in your hands, twirl them around and really get to know them. A complete set is shown in Figure 21.9. You can't help but fall in love with these symmetrical solids.

Figure 21.9

The Platonic solids are worth their weight in gold.

The Least You Need to Know

- A prism has two congruent, parallel, polygonal bases, with line segments connecting corresponding vertices. A pyramid has a polygonal base, with line segments connecting all vertices to a point.

- A cylinder is a generalization of a prism; a cone is a generalization of a pyramid; a sphere is a generalization of a circle.

- Euler's formula specifies the relationship between the vertices, edges, and faces of all polyhedra: $V + F = E + 2$.

- There are five Platonic solids: the tetrahedron, the octahedron, the icosahedron, the cube, and the dodecahedron.

Chapter 22

Under Construction

In This Chapter

- The tools of the trade
- Construction regulations
- How to bisect
- How to construct parallel and perpendicular lines

There have been several places throughout this book when I have constructed something. I've had two points and I needed the line segment between them, or I've had a point and a line and I've needed a parallel or perpendicular line. At those times I have virtually snapped my fingers and the line or line segment magically appeared.

Don't get me wrong. I haven't done anything illegal. I'm not confessing because the geometry police are knocking at my door, warrant in hand, ready to arrest me. I just wanted to give you the tools so that you wouldn't be completely lost if you had to make something (like a circle or a line segment), or do something (like find the midpoint of a segment or bisect an angle).

Tools of the Trade

There's no need to rush out to the store and buy lots of tools. You will really only need two things: a straight edge and a compass. Of course, a protractor comes in handy if you want to measure angles. But that's a luxury item, not a necessity.

Figure 22.1

Tools of the trade: a compass, a ruler, and a protractor.

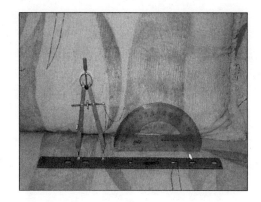

I just got my compass back from the shop—I upgraded the legs and put in a new pencil. I'm the envy of all the geometers in town with my sharp-looking compass! Not only that, but my straight edge rules!

Straight Edge

A straight edge is just what it sounds like. It is something straight. Remember that "straight" is one of those difficult-to-define-but-everybody-knows-what-it-means kind of terms.

When you think of a straight edge, you might immediately think of a ruler. A ruler is usually made of plastic or wood, and it has marks on both edges of one side. One set of marks indicates lengths in inches, and the other set of marks indicates centimeters.

> **Tangent Line**
>
> The United States is one of the last countries to insist on using the English system of measurement. Even the English have switched over to the metric system!

If you don't own a ruler, do not despair. There are other straight edges besides rulers. I personally keep a credit card or an ATM card in my wallet at all times. I don't use them to buy things, I use them when I need to draw a straight line. You would be surprised how many times I find myself in a situation where I need to get straight to the point!

Compass

A compass has two legs attached at one end, almost like two line segments that come together to form an angle. You can change the size of the angle that the two legs form. One leg usually has a sharp metal point, and the other leg usually has a pencil attached. You can use it to make circles or arcs (parts of circles). You can also use its sharp point to break the ice at parties.

To make a circle with a compass, you first need to decide how big you want the radius to be. If the two legs of the compass make a large angle, then the circle you draw will have a large radius. Place the sharp point where you want the center of the circle to be located. Then pivot the pencil-leg of the compass around the sharp, stationary leg. If you want to draw an arc, just pivot a little. If you want the whole circle, pivot all the way around.

Protractors

A protractor is a tool used to measure angles. When I formally introduced you to angles (way back in Chapter 4), I went into great detail about how to use a protractor to measure angles. Although a protractor is a great tool for measuring angles (not to mention the fact that it's our only tool for measuring angles—besides using our trigonometric ratios, that is), it can only measure angles that are either acute, right, or obtuse. In order to measure bigger angles, you will need to break them into smaller angles and use your protractor and the Angle Addition Postulate (Postulate 4.2).

You might be tempted to use the markings on your ruler and protractor to bisect line segments and angles. Do not give in to this temptation. Measuring devices are limited in their accuracy. You don't know how your ruler got its marks. The equipment might have malfunctioned on the day your tools were made (it's been known to happen!).

Tangent Line

The rules for using a compass and a straight edge were set down by the Greeks:
- ♦ You are allowed to use your straight edge to draw a line through two given points.
- ♦ You are allowed to draw a circle with a given center and radius with a compass.
- ♦ You are not allowed to use these instruments in any other way, and the straight edge cannot be graduated or marked.

Even if your ruler was manufactured properly, the best that it can do is enable you to locate the general vicinity of the midpoint of a line segment. But I will show you how to find the *exact* midpoint of a line segment. The same thing holds for bisecting angles: you might be able to measure an angle with a protractor, and divide that value in half and draw an angle with that new measure. But again, that is just an estimate. Your protractor is limited in its accuracy, and when you bisect an angle, you want it exact.

Bisection

Bisection is a tricky thing. Whenever you divide something into two equal parts, the parts must be exactly equal. There is no room for error. The two parts must measure up under the closest scrutiny. I'm as picky as a bank robber dividing up the loot: I want accuracy down to the penny.

Bisecting Segments

Your mission, should you decide to accept it, is to take the line segment $\overline{AB}$ shown in Figure 22.2 and bisect it using only a compass and a straight edge.

Figure 22.2

The line segment $\overline{AB}$.

A   B

This is not as daunting a task as getting the crew of the Apollo 13 home safely, but it's close. All you have is a compass, which is useful for drawing arcs or circles, and a straight edge, which is useful for drawing lines.

Try the following: Set your compass angle so that you can draw a circle with a radius between two-thirds and three-fourths of the length of $\overline{AB}$. The exact radius is not important, it just needs to be in this ballpark. Then draw two circles with your compass (keeping the radius the same—this is important!): the first one with the center at A, the second one with the center at B. Because the radius of your circle was big enough, these two circles should intersect twice: once above $\overline{AB}$ and again below $\overline{AB}$. Use your straight edge to connect these two points of intersection. The intersection of this new line segment with your original segment $\overline{AB}$ is the midpoint of $\overline{AB}$. I was constructing this with you (in spirit, at least), and I'll show you my drawing in Figure 22.3 so you can compare it with yours.

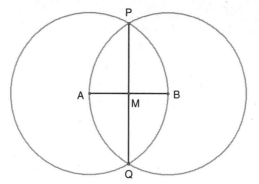

Figure 22.3

Constructing the midpoint of $\overline{AB}$ using circles and a line segment.

Of course, those kids on the playground don't believe me. Maybe you don't, either. Well, I'm not afraid to put my logic where my math is and walk you through the proof.

Because the circles have the same radius, $\overline{AP} \cong \overline{PB} \cong \overline{AQ} \cong \overline{BQ}$, quadrilateral APBQ has two pairs of opposite sides congruent, which makes it a parallelogram. Because $\overline{AB}$ and $\overline{PQ}$ are diagonals of this parallelogram, you know that they bisect each other. So the point of intersection of these two segments must be the midpoint of $\overline{AB}$.

Bisecting Angles

Suppose you have an angle, like $\angle ABC$ shown in Figure 22.4. How can you bisect this angle using only a straight edge and a compass?

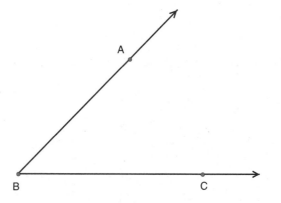

Figure 22.4

The focus of your next bisection: $\angle ABC$.

Think back to all of the proofs that involved isosceles triangles. The angle bisector, the median and the altitude all lie on the same line. So if B was the vertex angle of an isosceles triangle, all you would need to bisect this angle is to construct a ray that starts at B and passes through the midpoint of the base. And you can find the midpoint of a segment. Take a look at Figure 22.5.

Figure 22.5

Constructing the isosceles tri-
angle to find the midpoint of
the base and bisect ∠ABC.

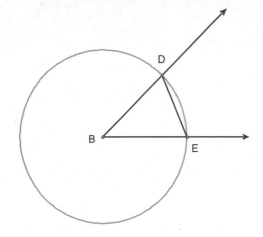

The only question that remains is where to get the isosceles triangle. Use your compass! Your compass is good for drawing circles, and if you center your circle at B, then the point where the circle and each ray of your angle intersect will help you create two radii: $\overline{BD}$ and $\overline{BE}$. Because the radii of a circle are congruent, ∆BDE is an isosceles triangle.

Let me walk you through the actual construction site. Set your compass to a fixed angle and draw a circle centered at B. Call the point of intersection of the circle and $\overrightarrow{BA}$ point D, and the point of intersection of the circle and $\overrightarrow{BC}$ point E. Use your compass again (without changing the angle ... you want to make circles with the same radius throughout this construction), draw two circles: the first with center D, the second with center E. Notice that these two circles intersect at two points: point B and another point. Call that second point of intersection F. Now use your straight edge to draw the ray $\overrightarrow{BF}$. This ray bisects ∠ABC.

When I walked you through the logic of the construction I began by constructing an isosceles triangle. The triangle itself was not important, since all I needed was the midpoint of the base. To understand the shortcuts I took, take a look at Figure 22.6. Because I used circles of the same radius throughout the construction, I was able to avoid the isosceles triangle construction and construct two triangles: ∆BDF and ∆BEF. These triangles are congruent by the SSS Postulate (Postulate 12.1). In this construction, $\overline{BD}$, $\overline{BE}$, $\overline{DF}$, and $\overline{EF}$ are all radii of congruent circles, so they are all congruent, and $\overline{BF}$ is congruent to itself. Thus by CPOCTAC, ∠DBF ≅ ∠EBF. That should convince the most skeptical person in the audience.

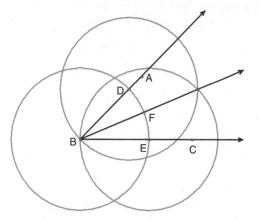

Figure 22.6

The construction steps used to bisect an angle.

Tangent Line

Bisecting angles can be done using two basic tools: the compass and the straight edge. You can divide an angle into four smaller, congruent angles by bisecting the bisected angles. You can divide an angle into eight smaller, congruent angles by bisecting again. But what about trisecting an angle using only your basic tools? This question bothered the Greeks, and it wasn't until 1837 that Pierre Wantzel proved that it was impossible to trisect an arbitrary angle using only a straight edge and a compass.

Constructing Lines

There have been times when you were given a line (or line segment) and a point (which might or might not be on the line or line segment), and you were instructed to construct a line parallel or perpendicular to that original line. At the time, I invoked either Euclid's 5th postulate for constructing a parallel line, or Theorem 10.1 for constructing a perpendicular line. Now I'll show you how. I'll start with the perpendicular and move into the parallel.

Perpendicular Lines

To construct a perpendicular line with only a straight edge and compass, you can apply the technique learned for constructing midpoints. That midpoint technique sure does come in handy! It's a good thing I talked about it first.

Suppose you have a point P and a line segment $\overline{AB}$, as shown in Figure 22.7.

Figure 22.7

A point P *and a line segment* $\overline{AB}$.

First, assume that the point is not on the line segment. Take your compass and make an angle large enough so that the radius of the circle it draws will be larger than the distance between the point and the line. Keep your compass setting the same throughout this construction. Use your compass to draw a circle centered at P that cuts $\overline{AB}$ twice. Label the points where the circle and the segment intersect C and D. Now use your compass to draw two congruent circles: one centered at C, the other centered at D. Those two circles will intersect twice: at P and at a point on the other side of $\overline{AB}$. Call the second point of intersection Q. Use your straight edge to draw the line (or line segment, depending on what you are asked to construct) $\overleftrightarrow{PQ}$: This is a line that passes through P and is perpendicular to $\overline{AB}$.

I can prove that $\overleftrightarrow{PQ} \perp \overline{AB}$ by examining the properties of the quadrilateral CPDQ shown in Figure 22.8. Because all four sides of the quadrilateral are congruent (they are radii of congruent circles), this quadrilateral is a rhombus. It is a well-known fact (established in Chapter 15) that the diagonals of a rhombus are perpendicular.

Figure 22.8

Quadrilateral CPDQ *generated in your construction.*

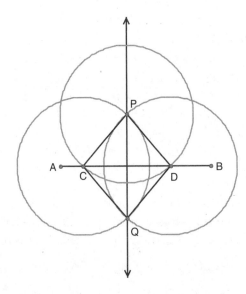

If you are given a line (or line segment) and a point on the line, there's not much difference in the construction. You still use your trusty compass to create your first circle, centered at the given point. You mark off the points where the circle and the line intersect, and then draw two more congruent circles, each one centered at an intersection point. Use a straight edge to join the points where the two circles intersect, and you have your perpendicular line!

Parallel Lines

Now that you know how to construct a perpendicular line when you are given a line and a point, you can construct a line passing through a particular point that is parallel to a specific line. In this case, the point must be off of the specific line.

Suppose you are given a point P and a line l, as shown in Figure 22.9. In order to construct a line that passes through P and is parallel to l, you need to first construct a line m that passes through P and is perpendicular to l. Then construct a line n that passes through P and is perpendicular to m. If you believe in Euclid's 5th postulate, lines l and n are parallel to each other.

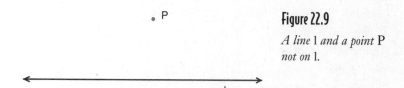

Figure 22.9

A line l and a point P not on l.

Circular Constructions: Off on Yet Another Tangent

It seems like everything you have constructed so far has used a circle in some form or another. Circles are the "cornerstones" of geometric constructions because of the congruent radii.

I have talked about lines that are tangent to circles, and now I'll show you how to draw one. Start with a circle centered at A, like the one shown in Figure 22.10, and a point P on the circle, where you want your tangent line to touch your circle. To construct the tangent line, the first thing you will need is a radius that passes through point P. Now that you have the line segment $\overline{AP}$, you can construct a line that passes through P and is perpendicular to $\overline{AP}$. That, my friend, is your tangent line.

Figure 22.10

A circle and a point P on the circle.

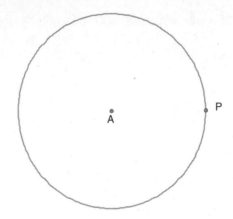

Constructing Quadrilaterals

Some quadrilaterals are easier to draw than others. To draw a generic, run-of-the-mill, nothing fancy quadrilateral, all you need is four points in a plane (or on a sheet of paper) to serve as the vertices of the quadrilateral. If you use your straight edge to draw the line segments between the consecutive vertices, a quadrilateral is born. I will go down the list of popular quadrilaterals, and summarize how to construct each type. Because mathematics is not a spectator sport, you'll need to follow along using the tools of the trade.

First I'll show you how to draw a trapezoid. Remember that a trapezoid has one set of parallel opposite sides, which are called the bases of the trapezoid. In order to draw a trapezoid, you need to specify three noncollinear points on the plane. Decide which two of these points will form one of the bases. Use your straight edge to draw the line segment connecting these two points. Now, take that line segment, and the third point (that is not part of the line segment) and construct a line passing through that third point that is parallel to the line segment. Choose any other point on that parallel line to be the fourth vertex of your quadrilateral. Connect consecutive vertices using your straight edge and you have a trapezoid.

Kites

It's time now to draw one of the most famous quadrilaterals: a kite. To draw a kite, start with a line segment $\overline{AB}$. Use your compass to draw two circles, one centered at A and the other centered at B. The two circles that you draw do not need to be congruent, but they do need to intersect off of $\overline{AB}$. Label the points of intersection C and D. Use your straight edge to connect the consecutive vertices. The resulting quadrilateral ACBD is a kite. This is easy to prove, because $\overline{AC}$ and $\overline{AD}$ are radii of one circle, and hence are congruent, and $\overline{BC}$ and $\overline{BD}$ are radii of the other circle, and are congruent. All it takes to make a kite is a quadrilateral with two sets of congruent adjacent sides. I have included all of the steps involved in kite construction in Figure 22.11.

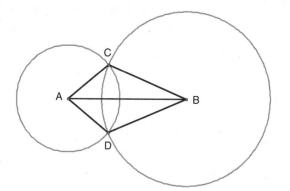

Figure 22.11

The geometric construction of a kite.

Parallelograms

To construct a parallelogram, start with three noncollinear points, A, B, and C, which will serve as three of the four vertices of your parallelogram. You will find the fourth vertex after constructing a couple of parallel lines. Use your straight edge to draw $\overline{AB}$. Construct the line parallel to $\overline{AB}$ that passes through C. Now use your straight edge to draw $\overline{AC}$. Construct the line that is parallel to $\overline{AC}$ and passes through B. The two lines that you constructed should intersect; label that point D. Quadrilateral ACDB is a parallelogram by definition, because $\overline{AB} \parallel \overline{CD}$ and $\overline{AC} \parallel \overline{BD}$.

Rectangles

To make a rectangle, you will need two points, A and B. Use your straight edge to construct $\overline{AB}$. Construct two lines that are perpendicular to $\overline{AB}$: one that passes through A, the other that passes through B. Choose a point C on one of these lines. Construct the line that is parallel to $\overline{AB}$ and passes through C. This line should intersect with both of the perpendicular lines you drew earlier: One point of intersection is C. Label the other point of intersection D. If C lies on the line perpendicular to $\overline{AB}$ that passes through A, then the quadrilateral ACDB is a rectangle. If C lies on the line perpendicular to $\overline{AB}$ that passes through B, then the quadrilateral ADCB is a rectangle.

Rhombuses

To construct a rhombus, start with one vertex, A. Use your compass to draw a circle centered at A. Choose two points that lie on the circle, and label them points B and C.

These two points will also be vertices of your rhombus. To find the fourth vertex, take your compass and make two more congruent circles (congruent to the first circle), one centered at B and the other centered at C. Those two circles should intersect twice. One point of intersection will be A. Call the other point of intersection D. Use your straight edge to construct the line segments $\overline{AB}$, $\overline{AC}$, $\overline{BD}$, and $\overline{CD}$. The resulting quadrilateral ABDC will be a rhombus. That is easy to prove, because all four sides of the quadrilateral are radii of congruent circles, and hence are congruent.

Squares

Last, but not least, is the construction of a square. Constructing a square is similar to constructing a rhombus. Start with a vertex A and use your compass to draw a circle centered at A. Choose one point that lies on the circle, and label the point B. Use your straight edge to construct $\overline{AB}$. Next, construct two lines perpendicular to $\overline{AB}$: one passing through point A, the other passing through point B. One of these lines will be tangent to the circle, and the other one will intersect the circle at two points. Choose one of those points, and label it C. Construct the line that is parallel to $\overline{AB}$ and passes through C. That line will intersect the tangent line you constructed earlier. Call the point of intersection D. Use a straight edge to construct the line segments $\overline{AC}$, $\overline{CD}$, and $\overline{BD}$. The resulting quadrilateral ACDB is a square.

That takes us through the constructions of many of the interesting geometric shapes. Any other constructions will use the same tools, and will probably build on what I've shown you. Remember, it's not enough just to read along while I do all the work. You need to be right there beside me, compass and straight edge in hand, laying your foundation and building high in the sky.

The Least You Need to Know

- The compass is instrumental in constructing congruent segments, and the straight edge is used to construct line segments in general.

- Bisection is performed by constructing congruent, intersecting circles.

- To construct a parallel line, construct one perpendicular line and then another.

- To construct a quadrilateral, first think of the properties of that quadrilateral, and then use circles, parallel lines, and perpendicular lines to make it.

When Geometry and Algebra Intersect

In This Chapter

- ◆ Calculating the distance between two points
- ◆ The slope of a line
- ◆ Slopes of parallel and perpendicular lines
- ◆ Graphing lines

You have explored the properties of a variety of geometrical objects throughout this book. You have learned about circles, parallel lines, perpendicular lines, triangles, and more! What you might not realize is that you have examined these shapes through a geometer's eyes.

There are other ways to look at these objects. Algebraists (the folks that study algebra) have eyes, too. They know all about these shapes, but their definitions are somewhat different than that of a geometer. In order for a geometer and an algebraist to have a meaningful conversation about, say, circles, they have to be sure that they are talking about the same thing.

An algebraist might describe a circle with center (0,0) and radius length r as "the set of all points (x,y) in the Cartesian coordinate system that satisfy the equation $x^2 + y^2 = r^2$." A geometer will describe a circle as "the set of all points a fixed length r from the center O." Before it comes to blows, someone needs to step in and check to see whether they are describing the same object. In this chapter, you will gain valuable training in algebraist/geometer mediation.

The Cartesian Coordinate System

Although geometers are quite content to have their points, lines, circles, and polygons floating around in space, algebraists are not. They require the position of an object to be specified clearly, so that they can find it with their eyes closed. So the Cartesian coordinate system was developed. This coordinate system involves dividing up a plane into four parts, called quadrants, using two perpendicular lines. One of the perpendicular lines is usually drawn horizontally, the other one vertically. They are given special names. The horizontal line is called the x-*axis* and the vertical line is called the y-*axis*. Their point of intersection is called the *origin*.

> **Tangent Line**
>
> The Cartesian coordinate system was named after the French philosopher René Descartes. He was instrumental in merging geometry and algebra together to form analytical geometry.

Each point in this coordinate system has an x-coordinate and a y-coordinate. These coordinates are written in the form (x,y), where x is the x-coordinate and y is the y-coordinate. The origin has the representation (0,0). Points to the right of the origin have a positive x-coordinate, and points to the left of the origin have a negative x-coordinate. Points above the origin have a positive y-coordinate, and points below the origin have a negative y-coordinate. Figure 23.1 shows the Cartesian coordinate system, and the point located at (1,2).

The Cartesian coordinate system divides the plane into four quadrants that are numbered I through IV. Quadrant I is located in the upper-right corner. Points in this quadrant have positive x-coordinates and positive y-coordinates. Quadrant II is in the upper-left corner. Points in this quadrant have negative x-coordinates and positive y-coordinates. Quadrant III is in the lower-left corner. Points in this quadrant have negative x-coordinates and negative y-coordinates. Quadrant IV is in the lower-right corner. Points in this quadrant have positive x-coordinates and negative y-coordinates.

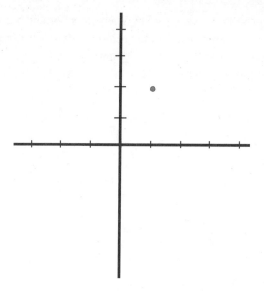

Figure 23.1

The Cartesian coordinate system and the point located at (1,2).

Finding Horizontal and Vertical Distances

One of the benefits of the Cartesian coordinate system is that finding distances is extremely straightforward. Horizontal and vertical distances are the easiest distances to find, and that is the natural place to start.

A horizontal segment is characterized by the property that the y-coordinate of the two endpoints of the segment is the same. For example, consider the horizontal segments with endpoints (2,4) and (6,4). Call the point (2,4) point P and the point (6,4) point Q. Now, the distance between these endpoints is just the length of $\overline{PQ}$. From an algebraic viewpoint, the distance between these endpoints is just the absolute value of the difference be-tween the x-coordinates (the difference between the y-coordinates is just zero, and that's a pretty boring distance). You have to take the absolute value because the distance between two points has to be positive. You can't say that two points are a negative three feet away from each other. So the distance between the two points is four units.

Eureka!

The distance between the points (x,y) and (z,y) is the absolute value of the difference between x and z, and is written as $|x - z|$. If P is the point (x,y) and Q is the point (z,y), then the distance between (x,y) and (z,y) is PQ.

You can find the length of a vertical line segment in a similar manner. With a vertical line segment the endpoints have the same x-coordinates. If a vertical line segment has endpoints P and Q, where P is the point (x,y) and Q is the point (x,w), then PQ, the distance between the points (x,y) and (x,w), is the absolute value of the difference in the y-coordinates: $|y - w|$.

What the Cartesian coordinate system enables you to do is place points in exact locations, and then use their coordinates to calculate distances between points or lengths of segments. Perhaps now is a good time to get some practice.

> **Example 1**: Find the distances between the following pairs of points: (2,3) and (8,3), and (4,6) and (4,15).

> **Solution**: The first couple of points have the same y-coordinates, so take the absolute value of the difference in the x-coordinates: $|2 - 8| = 6$. The second pair of points have the same x-coordinates, so take the absolute value of the difference in the y-coordinates: $|6 - 15| = 9$.

To find the lengths of line segments that are not vertical or horizontal, you need to bring out the big guns, and the most famous theorem of them all.

The Pythagorean Theorem Goes the Distance

Now that you know your way around the coordinate plane, and can calculate the length of vertical and horizontal line segments using the coordinates of the endpoints, you are ready for one of the relationships between geometry and algebra. The first connection between geometry and algebra is the distance formula.

In algebra, if you have two points, say (x,y) and (a,b), the distance d between them is given by the formula $d = \sqrt{(x-a)^2 + (y-b)^2}$. This looks like a pretty mysterious formula, until you look at distance through the eyes of a geometer. Take a look at the two generic points shown in Figure 23.2. If P is the point (a,b) and R is the point (x,y), then d is just PR. All you need to do is throw in the point (x,b) (call it Q). If you do that, then the points P, Q, and R form a right triangle. You can apply the Pythagorean Theorem to relate PR to PQ and QR: $PQ^2 + QR^2 = PR^2$. So what? Well, $\overline{PQ}$ is a horizontal line segment, and $\overline{QR}$ is a vertical line segment. You just spent the last section learning how to find the lengths of horizontal and vertical line segments. You know that $PQ = |x - a|$ and $QR = |y - b|$. Also, $PR = d$. If you substitute these values into the preceding equation, you have

> **Solid Facts**
>
> The **distance** d between two points having coordinates (x,y) and (a,b) is given by the formula $d = \sqrt{(x-a)^2 + (y-b)^2}$.

$(PQ)^2 + (QR)^2 = (PR)^2$

$\left(|x-a|\right)^2 + \left(|y-b|\right)^2 = d^2.$

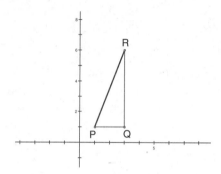

Figure 23.2

To find the distance between points P *and* R*, use* Q *to create a right triangle and use the Pythagorean Theorem.*

There is a little bit of simplification left. The absolute value of a number is always positive. Also, if you take any nonzero number and square it, you'll end up with something that is positive. So by taking the absolute value of x minus a and then squaring the result, you are being redundant: you do the subtraction, then change the sign if necessary (that's the absolute value part), and then square the result. Taking the absolute value of each term just adds unnecessary steps. You can eliminate the clutter and write the equation as $(x-a)^2 + (y-b)^2 = d^2$. All that is left is to apply the square

root property to solve for d: $d = \sqrt{(x-a)^2 + (y-b)^2}$. That's the mysterious distance formula. Now it is not so mysterious. It's just the algebraic version of the Pythagorean Theorem!

Example 2: Find the distance between the points (2,5) and (6,8).

Solution: Just apply the distance formula: $d = \sqrt{(6-2)^2 + (8-5)^2}$. Then $d = \sqrt{4^2 + 3^2} = \sqrt{25} = 5$ units.

The Midpoint Formula

When you think of the *midpoint* of a line segment, you will probably use a geometer's perspective. There are two endpoints and a point in the middle. The two endpoints are far away, and the midpoint is the best compromise, the happy medium. The midpoint doesn't favor either endpoint. Geometers aren't the only groups of mathematicians that describe midpoints. Algebraists are interested in these balancing points as well, and their definition involves giving a formula to calculate them. Here is an algebraist's definition of a midpoint: Given two points P and Q, with coordinates (a,b) and (c,d), respectively, the coordinates of the midpoint of the line segment $\overline{PQ}$ can be found using the formula

$\left(\dfrac{a+c}{2}, \dfrac{b+d}{2} \right).$

Notice that the x- and y-coordinates of the midpoint are just the averages of the x- and y-coordinates of the endpoints.

Solid Facts

The **midpoint** of the line segment with endpoints having coordinates (a,b) and (c,d) is given by the formula $\left(\dfrac{a+c}{2}, \dfrac{b+d}{2}\right)$.

Example 3: Find the midpoint of the line segment $\overline{PQ}$ where P is the point with coordinates (1,4) and Q is the point with coordinates (4,6).

Solution: Use the formula to find the average values of the x- and y-coordinates: The midpoint is located at the point with coordinates

$$\left(\frac{1+4}{2}, \frac{4+6}{2}\right) = (2.5, 5).$$

Finding Equations of Lines

Now that you know how to handle endpoints and midpoints of line segments in an algebraic as well as a geometric setting, it's time to expand your segments into lines. Armed with the equation of a line, you can easily find any point on that line.

An equation of a line will have two variables: an x variable and y variable. Notice that these variables match the coordinates used to specify the location of a point. This is not a coincidence; mathematicians do not play dice with the universe. If a line is vertical, its equation will only involve the x-coordinate. If a line is horizontal, its equation will only involve the y-coordinate. Otherwise, an equation of a line will involve both the x and y variables.

The two major components of a line are the slope and the intercepts. The slope of a line gives an indication of how steep (or flat, depending on your perspective) the line is. The intercepts indicate where the line crosses the x-axis or the y-axis. Remember that when you have two lines, only one of three things can happen: They are really the same line, they are parallel lines (and do not intersect), or they intersect at a point. The x-axis and the y-axis are lines in their own right. They are perpendicular to each other and intersect at the origin. If a line intersects the x-axis, the point of intersection is called the x-intercept. If a line intersects the y-axis, that point of intersection is called the y-intercept.

The Slope of a Line: Not Another Tangent!

The slope of a line is a measure of the steepness of the line. You can categorize lines as either vertical, horizontal, or inclined. The slope of a horizontal line is zero. Some people take that to mean that a horizontal line has no slope, but not mathematicians.

In English, it is all right to interchange zero and no. For example, if I have zero dollars in my pocket, I am tempted to say that I have no money. The two statements describe the same thing. The language of mathematics needs to be more precise. A line that has slope zero (or zero slope) is different than a line that has no slope. A horizontal line has zero slope whereas a vertical line is a line that has no slope. And if you need to be convinced that vertical lines and horizontal lines are different, just look at a drawbridge. Would you rather drive on the bridge when it is vertical or horizontal?

Tangent Line
The slope is the ratio of the change in the y-coordinates divided by the change in the x-coordinates. It is often referred to in calculus as a difference quotient. It involves the ratio (or quotient) of the change in the y-coordinates and the change in the x-coordinates (the difference). One of the many applications of calculus involves finding the slope of a line tangent to a curve at a specific point.

The *slope* of an inclined line can be thought of in terms of the ratio of the rise over the run. The rise is a measure of how much the y-coordinates change, and the run is a measure of how much the x-coordinates change. The slope is the ratio of these two measures. It can be found when you know any two points on the line. Suppose that a line passes through the points with coordinates (a,b) and (c,d). The slope of the line is calculated by taking a ratio of the change in the y-coordinates divided by the change in the x-coordinates. The equation used to find the slope of the line is

$$\frac{d-b}{c-a}.$$

It doesn't mater which point goes first. What is important is that you are consistent: Whichever point goes first in the numerator must also go first in the denominator.

> **Solid Facts**
>
> The **slope** of a line that passes through the points with coordinates (a,b) and (c,d) is given by the equation $\frac{d-b}{c-a}$.

Example 4: Find the slope of the line that passes through the points with coordinates (1,4) and (4,6).

Solution: Compute the ratio of the change in the y-coordinates divided by the change in the x-coordinates:

$$m = \frac{6-4}{4-1} = \frac{2}{3}$$

Tangent Line
If a line intersects with the x-axis, an angle is formed. The tangent of that angle will have the same numerical value as the slope of the line!

The Point-Slope Formula

The story so far: In the beginning, you have the coordinates of two points. Using these coordinates you can calculate the slope of the line that passes through them. That's the first step in coming up with the equation of the line.

To determine the equation of a line using the point-slope formula, you need two things: a point and a slope. You started out with two points, which appears to be more than what you need to use the point-slope formula. Because you also need the slope, you had to have that second point. After you have the slope, you can use whichever one of the two points is your favorite. Just as it didn't matter which point went first when calculating the slope, it doesn't matter which point you use in the point-slope formula. Either one will work.

Take your point, say (r,s), and your slope, say m, and substitute into the formula: $y - s = m(x - r)$. Distribute the m and collect all of your constants and terms involving x on one side, leave y by itself on the other. What you'll end up with is an equation that looks kind of like $y = mx + b$. This equation is referred to as the slope-intercept equation of a line.

> **Example 5**: Find the slope-intercept equation of the line which passes through the points with coordinates (2,3) and (3,6).

Eureka!

It doesn't matter which point goes first. Consistency is the key. In Example 4, the point (4,6) went first, consistently, in both the numerator and the denominator. If the point (1,4) had gone first you would have gotten the same result:

$$\frac{4-6}{1-4} = \frac{-2}{-3} = \frac{2}{3}.$$

Solution: You will want to use the point-slope formula, which requires you to find the slope of the line:

$$m = \frac{6-3}{3-2} = \frac{3}{1} = 3$$

You will also need a point. Either one will do. Use the point with coordinates (2,3). Substitute these pieces into the point-slope formula:

$$y - 3 = 3(x - 2)$$

Simplify this expression to get the final result:

$$y - 3 = 3x - 6$$

$$y = 3x - 3$$

The Intercepts

There are several ways that you can write the equation of a line. My personal favorite is the slope-intercept form. When a line is written in that form ($y = mx + b$), two interesting aspects of a line are revealed. In this form, the slope is readily available: It is the coefficient in front of the x variable. The other piece of information that is immediately available is the y-coordinate of the y-intercept: b. Recall that the y-intercept is the point where a line intersects with the y-axis. Any point on the y-axis has an x-coordinate equal to zero. So the line with equation y = mx + b has slope m and y-intercept (0,b).

> **Example 6**: Find the slope and the y-intercept of the line given by the equation y = 4x – 7.

> **Solution**: The slope is 4 and the y-intercept is the point (0,–7).

The Secret Lives of Parallel and Perpendicular Lines

When your line is in slope-intercept form, several observations can be made quickly and easily. For example, when two lines are parallel, they will not intersect. The only way that two lines will not intersect is if they have the same slope. For example, the line given by the equation y = 2x – 4 and the line given by y = 2x + 5 each have slope equal to 2. Because they have the same slope and different y-intercepts, they are parallel. Writing the equations for the lines in slope-intercept form made this observation almost immediate. Notice that the line given by y = 3x – 2 and the line given by y = 2x + 3 are not parallel, because they have different slopes. The first line has slope equal to 3, the second line has slope equal to 2.

Recognizing perpendicular lines is almost as easy. Two lines that are not vertical or horizontal are perpendicular if the products of their slopes equal –1. That means that if you have two perpendicular lines, one with slope m and the other with slope n, and you multiply the two slopes together, you will get –1. Always. For example, the line given by y = –2x + 1 and $y = \frac{1}{2}x + 3$ are perpendicular. The first line has slope –2 and the second line has slope $\frac{1}{2}$. The product of –2 and $\frac{1}{2}$ is –1, and that's what it takes to be perpendicular. The numbers –2 and $\frac{1}{2}$ are called negative reciprocals of each other.

Eureka!

Parallel lines have the same slope; perpendicular lines have slopes whose product is equal to –1. Two numbers whose product is equal to –1 are called negative reciprocals. Vertical and horizontal lines are perpendicular, but the slope of a vertical line is undefined and the slope of a horizontal line is 0. It makes no sense to multiply the slopes of these two lines together.

Graphing Lines: A Picture Is Worth a Thousand Words

If you are given the equation of a line and you are asked to come up with its graph, don't worry. Get out your handy straight edge and draw the Cartesian coordinate plane … you know, the grid. To draw a line, all you need is two points. Any two points that are on the line will do. The easiest way to find two points on a line is to pick two values for x and see what y has to be in each case. Then you have the coordinates of two points. Find those two points on your grid, and use your straight edge to connect them. There's your line.

Example 7: Graph the line given by the equation y = 2x + 1.

Solution: To graph this line, find two points. If x = 0 then y = 1 and you have the coordinates of one point: (0,1). If x = 1 then y = 3 and you have the coordinates of a second point: (1,3). Put these points on a grid and connect the dots with a straight edge, as shown in Figure 23.3.

Figure 23.3

The graph of the line given by y = 2x + 1.

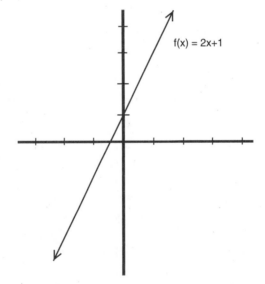

f(x) = 2x+1

Put Me in, Coach!

Here's your chance to shine. Remember that I am with you in spirit and have provided the answers to these questions in Appendix A.

1. Find the distance between the points (3,5) and (6,5).

2. Find the distance between the points (3,5) and (6,8).

3. Find the midpoint of the line segment $\overline{PQ}$ where P is the point with coordinates (3,5) and Q is the point with coordinates (6,8).

4. Find the slope of the line passing through the points with coordinates (3,5) and (6,8).

5. Find the slope-intercept equation of the line passing through the points with coordinates (3,5) and (6,8).

6. Find the slope and the y-intercept of the line given by the equation $y = 2x + 5$.

7. Are the following pairs of lines parallel, perpendicular or neither?

   a. $y = 3x + 7$ and $y = -3x - 7$

   b. $y = 3x + 7$ and $y = -\frac{1}{3}x - 7$

   c. $y = 3x + 7$ and $y = 3x - 7$

The Least You Need to Know

◆ The distance formula in algebra is just an algebraic interpretation of the Pythagorean Theorem.

◆ To find the midpoint of a line segment, just find the average values of the x- and y-coordinates of the endpoints.

◆ The slope of a line passing through two points is found by calculating the ratio of the change in the y-coordinates divided by the change in the x-coordinates.

◆ To find the equation of a line, you need a point, a slope, and the point-slope formula.

◆ If two lines are parallel, they have the same slope.

◆ The product of the slopes of perpendicular lines equals -1.

Chapter 24

Whose Geometry Is It Anyway?

In This Chapter

- ◆ Spherical geometry

- ◆ Hyperbolic geometry

- ◆ Taxi-cab geometry

- ◆ Max geometry

Euclidean geometry is sometimes referred to as plane geometry. Non-Euclidean geometry, on the other hand, refers to all of the other geometries. Euclidean and non-Euclidean geometry involve more than proving theorems. The ideas in Euclidean geometry can be used to derive formulas (like the Pythagorean Theorem, for example) or give explicit construction instructions (like how to construct an angle bisector). Using Euclidean geometry, you can estimate the height of a tree or the width of a lake. Unfortunately, these applications were established with the help of Euclid's 5th Postulate, and hold true only in Euclidean geometry. There are useful applications of non-Euclidean geometry as well; for example, navigation is one application of spherical geometry.

You might be wondering how Euclid's 5th Postulate was involved in the development of those applications. Most of the time you probably didn't see it. Euclid's 5th Postulate is sneaky, and one of its favorite places to hide is in the shadows of triangles. One reason that Euclid's 5th Postulate is able sneak around is because it was used to prove that the measures of the interior angles of a triangle add up to 180°. As a result, Theorem 11.1 became contaminated, and any theorem that makes use of a relationship between the interior angles of a triangle is suspect.

That's not to say that any time triangles are involved in a proof, the theorem only holds up in Euclidean geometry. Theorems can often be proven in more than one way. One way might involve triangles, another might involve something completely different. If a theorem can be proven in *at least one way* that isn't contaminated with the Parallel Postulate, then the theorem will hold in Euclidean and non-Euclidean geometry.

Non-Euclidean Geometry

Euclid had a hard time with the Parallel Postulate. As has already been mentioned, it is similar enough to the theorem about the existence and uniqueness of perpendicular lines to make a person think that the Parallel Postulate can be proven. Many brilliant mathematicians tried to prove the Parallel Postulate from Euclid's other postulates, and all have failed. It might be comforting to note that their failure was not a reflection of their ability as mathematicians. They were trying to do the impossible. Not just the impossible for their time, but the impossible for all time.

> **Tangent Line**
>
> There are several instances where mathematicians have proven that it is impossible to prove something. Although this concept might be difficult to understand and accept, it can be interpreted as permission to stop wasting time trying to prove a particular theorem.

Originally non-Euclidean geometry included only the geometries that contradicted Euclid's 5th Postulate. But then mathematicians realized that if interesting things happen when Euclid's 5th Postulate is tossed out, maybe interesting things happen if other postulates are contradicted. Each time a postulate was contradicted, a new non-Euclidean geometry was created. So the notion of non-Euclidean geometry had to be expanded. A non-Euclidean geometry is a geometry characterized by at least one contradiction of a Euclidean geometry postulate.

One of the reasons why non-Euclidean geometry is difficult to accept is that it goes against our practical experience. We perceive our world to be flat, even though the earth is spherical. It is easy to visualize a city as a grid with nonintersecting straight streets. That perception works because the curvature of the earth is insignificant when compared to the size of our cities. But non-Euclidean geometry has applications both in space and on our home planet.

Before we leave Euclid's world, it might be wise to remind yourself of the Parallel Postulate, first introduced in Chapter 5.

> **The Parallel Postulate.** Through a given point, not on a given line, only one parallel can be drawn to the given line.

You are at a point in the text when I need to be honest with you. Euclid is credited with being the father of geometry, but geometry has come a long way since Euclid's day. When you read current geometry books (like this one) it is easy to forget that Euclid wrote in Greek, using the language of his time. Although his writings might have been hip in his day, they lose a lot in the translation. There's nothing wrong with that. His writings served their purpose. The ideas he introduced in geometry have furthered development in many fields outside of mathematics, and geometry continues to develop even as I write. The point I am trying to make is that the wording of the definitions, theorems, and postulates in geometry has also changed with time, but its meaning has not. The phrasings of the definitions, theorems, and postulates in this book are equivalent to the ones that Euclid stated years ago, though they are not identical. I am taking a long time to confess my sin. Although I have credited this postulate to Euclid, the phrasing of it really belongs to John Playfair. It is equivalent to the one that Euclid came up with, but it is much more understandable.

Hyperbolic Geometry: Saddle Up!

Girolamo Saccheri (1667–1733) is credited with being the first to successfully study the logical consequences of a denial of the Parallel Postulate. Though he gets the credit, his is a sad tale. Saccheri tried to prove the Parallel Postulate by contradiction: He assumed that the Parallel Postulate was false, developed the resulting consequences, and looked for a contradiction. Much to his disdain, he was unable to establish a contradiction of any definition, theorem, or postulate. He did manage to violate several intuitive ideas, but a proof by contradiction must involve a violation of a postulate, theorem, or definition, not intuition. If he had admitted that his reasoning did not lead to a contradiction (and had been willing to explore the resulting paradoxical ideas), he would have anticipated the discovery of one form of non-Euclidean geometry by at least 100 years. This non-Euclidean geometry might have even been named in his honor. As it happened, he downplayed his results and concluded that, based on his work, the Parallel Postulate must be true. As a result of his incorrect conclusion, his results were not widely read and did not have much influence on the mathematical community until almost 150 years after his death.

Tangent Line

The denial of the Parallel Postulate leads to two alternatives:

◆ Given a line and a point not on the line, there are no lines through the point parallel to the original line. In other words, parallel lines do not exist.

◆ Given a line and a point not on the line, there are at least two lines through the point parallel to the original line.

Spherical geometry originates from the first alternative, hyperbolic geometry from the second.

It is ironic that Saccheri thought he was helping Euclidean geometry by trying to prove the Parallel Postulate. Had he delved into these paradoxical results, he would have been able to put to rest any doubts about the Parallel Postulate. He would have shown that the Parallel Postulate was in fact independent of Euclid's other postulates. That result would have helped Euclidean geometry more than anything.

This is all wishful thinking, though, because Saccheri did not have confidence in his results and he drew the wrong conclusion from his work. As it stands, Janos Bolyai (1802–1860) and Nikolai Lobachevsky (1793–1856) are the two mathematicians credited with the development of a logically consistent geometry based on a denial of the Parallel Postulate. As is typical, they never received full recognition of the value of their work during their lifetimes. This particular branch of non-Euclidean geometry is called hyperbolic geometry, or saddle-point geometry. The postulate that generates hyperbolic geometry is the Lobachevskian Postulate.

Tangent Line

Janos Bolyai published his results in an appendix in his father's book that was appropriately titled (in translation) *An Attempt to Introduce Studious Youth to the Elements of Pure Mathematics*. A copy of his work was also sent to the highly celebrated mathematician Carl Gauss. Gauss had already worked out the ideas written by Janos Bolyai, but had never published them. In a letter that he wrote to Janos's father, he mentioned that he was happy that these results would be published and that he was spared the trouble of writing them up himself. Rather than interpreting Gauss' comment as a compliment, Janos thought Gauss was trying to take credit for his work. Janos Bolyai lost interest in mathematics and never published again.

Lobachevskian Postulate: Through a point not on a line, there are infinitely many lines parallel to the given line.

In this geometry, several anti-intuitive results follow.

♦ The sum of the measures of the interior angles of a triangle is less than 180°.

♦ The sum of the measures of the interior angles of a triangle increase as the area decreases.

♦ Rectangles do not exist.

♦ Two triangles with congruent corresponding angles are congruent.

♦ There is an upper limit on the area of a triangle.

This is only the tip of the iceberg. Believe me when I say that strange things happen in hyperbolic geometry. Even stranger is the fact that a large group of scientists believe that the shape of the universe is best modeled using hyperbolic geometry!

Spherical Geometry

In Euclidean geometry, two-dimensional construction occurs within a plane. Recall that if you are given any three noncollinear points, there exists a plane that contains them. It should not surprise you that with spherical geometry (or elliptic geometry), everything is done on a sphere. Lines are defined as the great circles that encompass the sphere. A great circle is a circle whose center lies at the center of the sphere, as shown in Figure 24.1. No matter how they are drawn, each pair of great circles will always intersect. As a result, parallel lines do not exist.

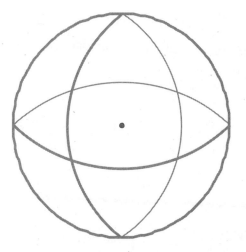

Figure 24.1

A model of spherical geometry.

The mathematician Bernhard Riemann (1826–1866) is credited with the development of spherical geometry. Ironically enough, he was born about the same time that hyperbolic geometry was developed by Bolyai and Lobachevsky, and he was instrumental in convincing the mathematical world of the merits of non-Euclidean geometry.

Although hyperbolic geometry needed only to contradict the Parallel Postulate, spherical geometry doesn't get off so easy. There are three changes in Euclid's axiomatic system necessary to successfully create spherical geometry. Remember that lines in Euclidean geometry are the great circles in spherical geometry.

> **Tangent Line**
>
> Spherical geometry is important in navigation, because the shortest distance between two points on a sphere is the path along a great circle.

Riemannian Postulate: Given a line and a point not on the line, every line passing though the point intersects the line. (There are no parallel lines).

All lines have the same finite length. (All great circles have a finite length: the circumference of the circles they describe.)

All perpendiculars to a straight line must meet at the same point.

This third postulate goes against our Euclidean intuition that if two lines are perpendicular to a third line, then the two perpendiculars are parallel. In spherical geometry, because there are no parallel lines, these two perpendiculars must intersect. But there is something more subtle involved in this third postulate. *All* perpendiculars meet at the same point. In this geometry, several counter-intuitive results follow. Some of these results are the same as the results in hyperbolic geometry, but for different reasons.

♦ The sum of the measures of the interior angles of a triangle is greater than 180°.

♦ All straight lines are bounded and finite in length.

♦ Rectangles do not exist.

♦ Two triangles with congruent corresponding angles are congruent.

♦ There is an upper bound on the area of a triangle.

♦ The intersection of two distinct lines is always two points.

Taxi-Cab Geometry

Taxi-cab geometry does not owe its existence to contradicting Euclid's 5th Postulate. Instead, this geometry has its origins in mathematicians using different methods to measure distances. The following story is one possible scenario for how taxi-cab geometry came into existence.

The Joint Meetings of the American Mathematical Society and the Mathematics Association of America has been scheduled to occur in New York City. Mathematicians arrive in droves, each of them taking a cab from the airport to their hotel. The cab fare they pay depends on the distance traveled. One financially conservative mathematician looks at a tourist map he picked up at the airport and calculates the distance between the airport and the hotel to be 12.6 miles. The rates posted on the window of the cab indicate that the fare charged is $1.00 for the first tenth of a mile and twenty cents for every tenth of a mile thereafter. After a quick calculation, the mathematician expects the total fare to be $26.00 (not including tip).

The mathematician gets $30.00 out to cover the cab ride and the tip. He is shocked when the cabby tells him the fare is $48.00. This cannot be! He calculated the distance between the airport and the hotel using Euclid's notion of distance (as interpreted in the Cartesian Coordinate System) to be 12.6 miles. Even if the map is a bit inaccurate, the difference between Euclid's notion of distance and the cabby's notion of distance are like night and day (at least to this financially conservative mathematician).

Being a good citizen, the mathematician pays the cabby what is owed, but throughout the conference he tries to understand why his estimate of the fare was so far off base. Toward the end of the conference, he finally understands that he measured the distance between the airport and the hotel "as the crow flies," whereas the cab must stick to the streets. The mathematician quickly establishes the rules for calculating distance in the taxi-cab world, and taxi-cab geometry is formed.

However taxi-cab geometry came about, it is interesting to note that if you redefine distance, you redefine the geometrical world. You can calculate distances in the taxi-cab geometry easily if you put your map on a Cartesian Coordinate System. Suppose you have two points, one with coordinates (1,3) and the other with coordinates (4,7), as shown in Figure 24.2. The distance between these points using the old-fashioned distance formula is found to be

$$d_{Euclid} = \sqrt{(4-1)^2 + (7-3)^2}$$
$$d_{Euclid} = \sqrt{3^2 + 4^2}$$
$$d_{Euclid} = \sqrt{9+16}$$
$$d_{Euclid} = \sqrt{25}$$
$$d_{Euclid} = 5$$

Figure 24.2

The Cartesian Coordinate System with points having coordinates (1,3) and (4,7).

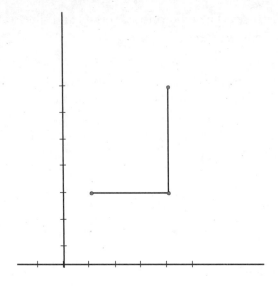

Eureka!

The taxi-cab distance between two points (a,b) and (c,d) is determined by the formula $d_{Taxi} = |a - c| + |b - d|$.

If you measure the distance between the two points using the taxi-cab metric, you will get $d_{Taxi} = 3 + 4 = 7$.

Mathematicians usually write the taxi-cab distance formula as $d_{Taxi} = |a - c| + |b - d|$.

Max Geometry

With the advent of the taxi-cab geometry, mathematicians learned that if they used a different method to measure the distance between two points, a new geometry was created. So they began to play around with other acceptable definitions of distance. Now there are more non-Euclidean geometries than you can shake a stick at.

Another interesting non-Euclidean geometry that has its origins in a new distance formula is the max geometry. In the max geometry, the notion of the distance between two points, say (a,b) and (c,d), is defined to be either $|a - c|$ or $|b - d|$, depending on which is bigger. That's how it got its name: the max distance is the maximum of $|a - c|$ and $|b - d|$. We usually write this distance formula as $d_{Max} = Max\{|a - c|, |b - d|\}$.

Let's go back to the last problem: You have two points, one with coordinates (1,3) and the other with coordinates (4,7), and you want to know the max distance between these two points. Using the formula,

$$d_{Max} = Max\{|4 - 1|, |7 - 3|\} = Max\{3, 4\} = 4.$$

So the distance between the points with coordinates (1,3) and (4,7) depends on what is meant by distance. The distance between them "as the crow flies" (that is, the Euclidean notion of distance) is 5. The distance between the two points according to the taxi-cab metric is 7, and the distance according to the max metric is 4. All three notions of distance have their uses, though the max metric is the most obscure, and is probably used by mathematicians more than any other group. It might be helpful to think of the max distance between two points as being the furthest distance in either the North/South or East/West directions.

Eureka!

The max distance between two points (a,b) and (c,d) is determined by the formula

$$d_{Max} = Max\{|a - c|, |b - d|\}.$$

How Many Shapes Can a Circle Have?

You have already learned that a circle is the set of all points a fixed distance away from one specific point that is called the center of the circle. The notion of a circle depends heavily on what is meant by distance. If you change the way that you measure distance, you change the way that you draw a circle.

Using the familiar definition of distance, a circle has the shape you have come to associate with a circle. But a "circle" (by definition) has a different shape in the taxi-cab and max geometries. I will discuss the shape of a circle in these other two geometries, but please use this information wisely. If you are told to arrange the chairs in a room in the shape of a circle, use a Euclidean circle rather than a taxi-cab circle!

Let me remind you of what the unit circle looks like in Euclidean geometry (in the Cartesian Coordinate System), with the center of the circle located at the origin. The unit circle is the set of all points 1 unit away from the origin. Certainly the points (1,0), (0,1), (–1,0) and (0,–1) are all on the unit circle, because their Euclidean distance from the origin is 1 unit. But the following points are also on the unit circle

$$\left(\tfrac{\sqrt{2}}{2}, \tfrac{\sqrt{2}}{2}\right), \left(\tfrac{\sqrt{2}}{2}, -\tfrac{\sqrt{2}}{2}\right), \left(-\tfrac{\sqrt{2}}{2}, \tfrac{\sqrt{2}}{2}\right) \text{ and } \left(-\tfrac{\sqrt{2}}{2}, -\tfrac{\sqrt{2}}{2}\right)$$

And there are many more. Points like $\left(\tfrac{1}{2}, \tfrac{\sqrt{3}}{2}\right)$ are also on the unit circle. Any point (a,b), where $a^2 + b^2 = 1$ lies on the Euclidean unit circle. Figure 24.3 shows the Euclidean unit circle. There are no surprises.

Figure 24.3

The Euclidean unit circle.

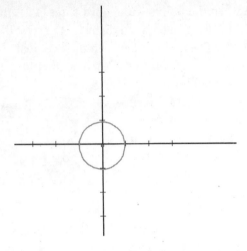

Open your mind a bit, and let's try and visualize the taxi-cab unit circle centered at the origin. Think about the points that are 1 unit away from the origin using this new notion of distance. You could still travel to the points (1,0), (0,1), (–1,0) and (0,–1). So these points are on the Euclidean unit circle as well as the taxi-cab unit circle. Suppose now that you travel ½ unit to the right. How far up could you travel so that the total distance that the cab took you was 1 unit? You could travel up ½ unit, so that the total distance traveled was ½ + ½ = 1 unit. This is true whether you travel left or right first, and then up or down second. So the points $\left(\frac{1}{2},\frac{1}{2}\right)$, $\left(\frac{1}{2},-\frac{1}{2}\right)$, $\left(-\frac{1}{2},\frac{1}{2}\right)$ and $\left(-\frac{1}{2},-\frac{1}{2}\right)$ are also on the taxi-cab unit circle. What if you traveled to the right ¼ of a unit? Well, you could then travel ¾ of a unit up or down and you would have traveled a total distance of ¼ + ¾ = 1 unit as well. So any point (a,b), where $|a| + |b| = 1$, lies on the taxi-cab unit circle. I have drawn the taxi-cab unit circle for you in Figure 24.4.

Figure 24.4

The taxi-cab unit circle.

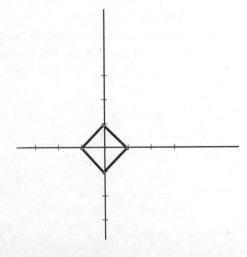

Notice that the resulting shape has some interesting properties. First of all, it is a quadrilateral. The next thing to observe is that the diagonals of this quadrilateral each have length 2, so they are congruent. Notice that the diagonals bisect each other. Finally, the opposite sides are congruent. Hey, this quadrilateral is a parallelogram, a rectangle, and a rhombus! It's a square! The taxi-cab unit circle has the shape of a Euclidean square! You can find the Euclidean length of the sides of this square: The Euclidean length is $\sqrt{2}$. So the taxi-cab unit circle is a Euclidean square with sides having length $\sqrt{2}$.

Finally, let's explore what a max unit circle centered at the origin looks like. Again, the points (1,0), (0,1), (–1,0) and (0,–1) are all on the max unit circle, because the maximum of the North/South and East/West distances is 1 unit. What about the point $\left(1,\frac{1}{2}\right)$? Is that point on the Max unit circle? Well, the Max distance between this point and the origin is $d_{Max} = Max\left\{1,\frac{1}{2}\right\} = 1$, so the point $\left(1,\frac{1}{2}\right)$ is on the max unit circle. What about the point $\left(1,\frac{3}{4}\right)$? It's there as well. In fact, any point (a,b), where $Max\{|a|,|b|\} = 1$ is on the max unit circle. So the point $\left(-\frac{1}{2},1\right)$ is on the Max unit circle. I have drawn the max unit circle for you in Figure 24.5.

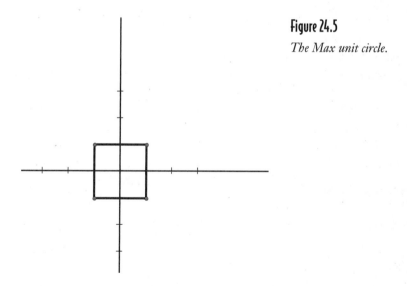

Figure 24.5

The Max unit circle.

Notice that the resulting shape has some interesting properties. First of all, the lines make up a quadrilateral. The next thing to observe is that the sides of this quadrilateral each have length 2, so they are congruent. Notice that two of the sides are horizontal lines, and the other two sides are vertical lines. So the sides of the quadrilateral are perpendicular, and they form a right angle. Hey, this quadrilateral is a parallelogram, a rectangle, and a rhombus! It's a square! The max unit circle has the shape of a

Eureka!

The taxi-cab unit circle and the max unit circle both have the shape of a Euclidean square! The difference between the two "unit circles" is the length of the sides of the squares.

Euclidean square, just like the taxi-cab unit circle was a square. But the max unit circle is a Euclidean square with side length 2, whereas the taxi-cab unit circle was a Euclidean square with sides having length $\sqrt{2}$. So the taxi-cab unit circle and the max unit circle are both Euclidean squares with different lengths.

The Least You Need to Know

- Non-Euclidean geometries got their start by contradicting Euclid's Parallel Postulate.

- Hyperbolic geometry assumes that given a line and a point not on the line, there exists infinitely many lines passing through the point which are parallel to the given line.

- Spherical geometry assumes that parallel lines do not exist, lines have finite length, and that all lines perpendicular to a given line intersect at the North and South Poles.

- Euclidean geometry measures distance "as the crow flies" whereas taxi-cab geometry and the max geometry have different notions of distance.

- The shape of a circle depends on the method used to measure distance.

Chapter 25

Transformations

In This Chapter

- ◆ What are isometries?
- ◆ The four basic types of isometries
- ◆ Dilations
- ◆ Symmetry

There is very little in our world that is static. Living things like to move. I remember thinking how great it would be to move out of my parents' house. Looking back, I didn't realize that I had it so good; moving into my own house meant that I had to pay for rent, utilities, and food!

Geometrical figures like to move as well. They can slide along the plane, turn around, or flip upside down. They can do just one of these moves or they can do several of them in succession, depending on their ambitions. Throughout this process, however, they will keep the same size and shape. Triangles won't change into quadrilaterals or change angle measurements. They won't grow or shrink either.

That's not to say that geometric figures have to stay the same size throughout their lives. They *can* grow or shrink. But when they do, only their sides are allowed to change; their angle measures cannot. That means that as geometrical figures grow or shrink, they will remain similar to what they were before.

Isometries

The word isometry is used to describe the process of moving a geometric object from one place to another without changing its size or shape. Imagine two ants sitting on a triangle while you move it from one location to another. The location of the ants will change relative to the plane (because they are on the triangle and the triangle has moved). But the location of the ants relative to each other has not. Whenever you transform a geometric figure so that the relative distance between any two points has not changed, that transformation is called an *isometry*. There are many ways to move two-dimensional figures around a plane, but there are only four types of isometries possible: translation, reflection, rotation, and glide reflection. These transformations are also known as rigid motion. The four types of rigid motion (translation, reflection, rotation, and glide reflection) are called the basic rigid motions in the plane. These will be discussed in more detail as the chapter progresses.

> **Tangent Line**
>
> For three-dimensional objects in space there are only six possible types of rigid motion: translation, reflection, rotation, glide reflection, rotary reflection, and screw displacement. These isometries are called the basic rigid motions in space.

> **Solid Facts**
>
> An **isometry** is a transformation that preserves the relative distance between points.
>
> Under an isometry, the **image** of a point is its final position.
>
> A **fixed point** of an isometry is a point that is its own image under the isometry.

An isometry in the plane moves each point from its starting position P to an ending position P′, called the *image* of P. It is possible for a point to end up where it started. In this case P = P′ and P is called a *fixed point* of the isometry. In studying isometries, the only things that are important are the starting and ending positions. It doesn't matter what happens in between.

Consider the following example: suppose you have a quarter sitting on your dresser. In the morning you pick it up and put it in your pocket. You go to school, hang out at the mall, flip it to see who gets the ball first in a game of touch football, return home exhausted and put it back on your dresser. Although your quarter has had the adventure of a lifetime, the net result is not very impressive; it started its day on the dresser and ended its day on the dresser. Oh sure, it might have ended up in a different place on the dresser, and it might be heads up instead of tails up, but other than those minor differences it's not much better off than it was at the beginning of the day. From the quarter's perspective there was an easier way to end up where it did. The same effect could have been accomplished by moving the quarter to its new position first thing in the morning. Then it could have had the whole day to sit on the dresser and contemplate life, the universe, and everything.

If two isometries have the same net effect they are considered to be equivalent isometries. With isometries, the "ends" are all that matters, the "means" don't mean a thing.

An isometry can't change a geometric figure too much. An isometry will not change the size or shape of a figure. I can phrase this in more precise mathematical language. The image of an object under an isometry is a congruent object. An isometry will not affect collinearity of points, nor will it affect relative position of points. In other words, if three points are collinear before an isometry is applied, they will be collinear afterwards as well. The same holds for between-ness. If a point is between two other points before an isometry is applied, it will remain between the two other points afterward. If a property doesn't change during a transformation, that property is said to be *invariant*. Collinearity and between-ness are invariant under an isometry. Angle measure is also invariant under an isometry.

If you have two congruent triangles situated in the same plane, it turns out that there exists an isometry (or sequence of isometries) that transforms one triangle into the other. So all congruent triangles stem from one triangle and the isometries that move it around in the plane.

Solid Facts _____

A property is **invariant** if it remains unchanged by a transformation.

You might be tempted to think that in order to understand the effects of an isometry on a figure, you would need to know where every point in the figure is moved. That would be too complicated. It turns out that you only need to know where a few points go in order to know where all of the points go. How many points is "a few" depends on the type of motion. With translations, for example, you only need to know the initial and final positions of one point. That's because where one point goes, the rest follow, so to speak. With isometries, the distance between points has to stay the same, so they are all kind of stuck together.

Because you will be focusing on the starting and ending locations of points, it is best to couch this discussion in the Cartesian Coordinate System. That's because the Cartesian Coordinate System makes it easy to keep track of the location of points in the plane.

Translations

When you translate an object in the plane, you slide it around. A *translation* in the plane is an isometry that moves every point in the plane a fixed distance in a fixed direction. You don't flip it, turn it, twist it, or bop it. In fact, with translations if you know where one point goes you know where they all go.

Solid Facts

A **translation** in the plane is an isometry that moves every point in the plane a fixed distance in a fixed direction.

The simplest translation is the "do nothing" translation. This is often referred to as the identity transformation, and is denoted I. Your figure ends up where it started. All points end up where they started, so all points are fixed points. The identity translation is the only translation with fixed points. With every other translation, if you move one point, you've moved them all. Figure 25.1 shows the translation of a triangle.

Figure 25.1

The translation of a triangle.

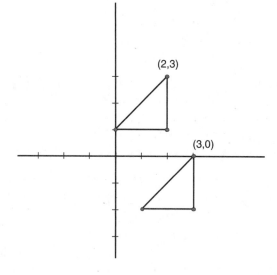

Translations preserve orientation: Left stays left, right stays right, top stays top and bottom stays bottom. Isometries that preserve orientations are called proper isometries.

Reflections

A *reflection* in the plane moves an object into a new position that is a mirror image of the original position. The mirror is a line, called the axis of reflection. If you know the axis of reflection, you know everything there is to know about the isometry.

Solid Facts

A **reflection** in the plane moves an object into a new position that is a mirror image of the original position.

Reflections are tricky because the frame of reference changes. Left can become right and top can become bottom, depending on the axis of reflection. The orientation changes in a reflection:

Clockwise becomes counterclockwise, and vice versa. Because reflections change the orientation, they are called improper isometries. It is easy to become disorientated by a reflection, as anyone who has wandered through a house of mirrors can attest to. Figure 25.2 shows the reflection of a triangle.

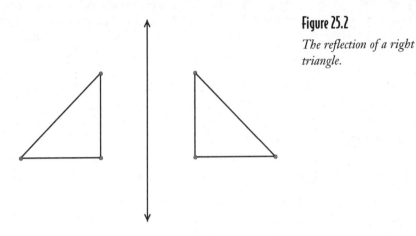

Figure 25.2

The reflection of a right triangle.

There is no identity reflection. In other words, there is no reflection that leaves every point on the plane unchanged. Notice that in a reflection all points on the axis of reflection do not move. That's where the fixed points are. There are several options regarding the number of fixed points. There can be no fixed points, a few (any finite number) fixed points, or infinitely many fixed points. It all depends on the object being reflected and the location of the axis of reflection. Figure 25.3 shows the reflection of several geometric figures. In the first figure, there are no fixed points. In the second figure there are two fixed points, and in the third figure there are infinitely many fixed points.

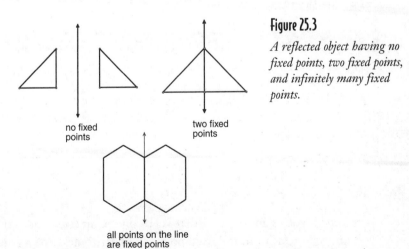

Figure 25.3

A reflected object having no fixed points, two fixed points, and infinitely many fixed points.

no fixed points

two fixed points

all points on the line are fixed points

> **CAUTION**
>
> **Tangled Knot** _____
>
> In Figure 25.3, you must be careful in the second drawing. Because of the symmetry of the triangle and the location of the axis of reflection, it might appear that all of the points are fixed points. But only the points where the triangle and the axis of reflection intersect are fixed. Even though the overall figure doesn't change upon reflection, the points that are not on the axis of reflection do change position.

A reflection can be described by how it changes a point P that is not on the axis of reflection. If you have a point P and the axis of reflection, construct a line l perpendicular to the axis of reflection that passes through P. Call the point of intersection of the two perpendicular lines M. Construct a circle centered at M which passes through P. This circle will intersect l at another point beside P, say P′. That new point is where P is moved by the reflection. Notice that this reflection will also move P′ over to P.

That's just half of what you can do. If you have a point P and you know the point P′ where the reflection moves P to, then you can find the axis of reflection. The preceding construction discussion gives it away. The axis of reflection is just the perpendicular bisector of the line segment $\overline{PP'}$! And you know all about constructing perpendicular bisectors.

What happens when you reflect an object twice across the same axis of reflection? The constructions discussed above should shed some light on this matter. If P and P′ switch places, and then switch places again, everything is back to square one. To the untrained eye, nothing has changed. This is the identity transformation I that was mentioned with translations. So even though there is no reflection identity per se, if you reflect twice about the same axis of reflection you have generated the identity transformation.

> **Tangent Line**
>
> Motion usually involves change. If something is stationary, is it moving? Should the identity transformation be considered a rigid motion? If you go on vacation and then return home, have you actually moved? Should the focus be on the process or the result? Using the term "isometry" rather than "rigid motion" effectively moves the focus away from the connotations associated with the "motion" aspect of a rigid motion.

Rotations

A _rotation_ involves an isometry that keeps one point fixed and moves all other points a certain angle relative to the fixed point. In order to describe a rotation, you have to know the pivot point, called the center of the rotation. You also have to know the amount of rotation. This is specified by an angle and a direction. For example, you could rotate a figure about a point P by an angle of 90°, but you need to know if the rotation is clockwise or counterclockwise. Figure 25.4 shows some examples of rotations about some points.

Solid Facts

A **rotation** is an isometry that moves each point a fixed angle relative to a central point.

Figure 25.4

Examples of rotations of figures.

Other than the identity rotation, rotations have one fixed point: the center of rotation. If you turn a point around, you don't change it, because it has no size to speak of. Also, a rotation preserves orientation. Everything rotates by the same angle, in the same direction, so left stays left and right stays right. Rotations are proper isometries. Because rotations are proper isometries and reflections are improper isometries, a rotation can never be equivalent to a reflection.

In order to describe a rotation, you need to specify more information than one point's origin and destination. Infinitely many rotations, each with a distinct center of rotation, will take a specific point P to its final location P′. All of these different rotations have something in common. The centers of rotation are all on the perpendicular bisector of the line segment $\overline{PP'}$. In order to nail down the description of a rotation, you need to know how two points change, but not just any two points. The perpendicular bisectors of the line segments connecting the initial and final locations of the points must be distinct. Suppose you know that P moves to P′ and Q moves to Q′, with the perpendicular bisector of $\overline{PP'}$ distinct from the perpendicular bisector of $\overline{QQ'}$. Then the rotation is specified completely. Figure 25.5 will help you visualize what I am trying to describe.

Eureka!

Rotation by 360° leaves everything unchanged; you've gone "full circle." You have seen three different ways to effectively leave things alone: the "do nothing" translation, reflection twice about the same axis of reflection, and rotation by 360°. Each of these isometries is equivalent, because the net result is the same.

The center of rotation must lie on the perpendicular bisectors of both $\overline{PP'}$ and $\overline{QQ'}$, and you know that two distinct nonparallel lines intersect at a point. The point of intersection of the perpendicular bisectors will be the center of rotation, C. To find the angle of rotation, just find m∠PCP'.

Figure 25.5

A rotation with center of rotation point C *and angle of rotation* m∠PCP'.

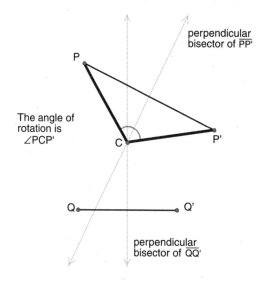

perpendicular bisector of $\overline{PP'}$

The angle of rotation is ∠PCP'

perpendicular bisector of $\overline{QQ'}$

Glide Reflections

A *glide reflection* consists of a translation followed by a reflection. The axis of reflection must be parallel to the direction of the translation. Figure 25.6 shows a figure transformed by a glide reflection. Notice that the direction of translation and the axis of reflection are parallel.

Solid Facts

A **glide reflection** is an isometry that consists of a translation followed by a reflection.

Notice that the orientation has changed. If you list the vertices of the triangle clockwise, the order is A, B, and C. If you list the vertices of the resulting triangle clockwise, the order is A', C', and B'. Because the orientation has changed, glide reflections are improper isometries.

In order to understand the effects of a glide reflection you need more information than where just one point ends up. Just as you saw with rotation, you need to know where two points end up. Because the translation and the axis of reflection are parallel, it is easy to determine the axis of reflection when you know how two points are moved. If P is moved to P' and Q is moved to Q', the axis of reflection is the line segment that connects the midpoints of the segments $\overline{PP'}$ and $\overline{QQ'}$. When the axis of reflection is known, you need to reflect the point P' across the axis of reflection. That will give you an intermediate point P*. The translation part of the glide reflection (in other words, the glide part) is the translation that moved P to P*. Now you know the translation and the axis of reflection, so you know everything about the isometry.

Because a glide reflection is a translation and a reflection, it will have no fixed points (assuming the translation is not the identity!). That's because nontrivial translations have no fixed points.

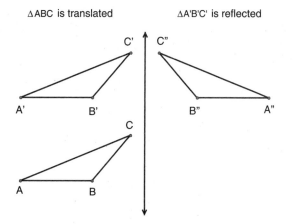

△ABC is translated   △A'B'C' is reflected

Figure 25.6

△ABC *undergoes a glide reflection.*

Dilations

A *dilation* is a transformation that changes the length of all line segments by the same proportion. A dilation does not change the shape of a figure, but it can change the size. Because the size of the figure changes, dilations are not isometries. Because the shape is invariant under a dilation, the original figure and its image are similar. By "similar" I mean the mathematical notion of similarity: Angles are congruent and sides are proportional.

If the proportion involved in the dilation is equal to one, then no noticeable change will have occurred. That would be the identity dilation. If the proportion is less than one, the figure will shrink. If the proportion is greater than one, the figure will grow.

Solid Facts

A **dilation** is a transformation that changes the length of all line segments by the same proportion.

Just as you saw with isometries, collinearity, between-ness and angle measures are invariant under a dilation. You saw that any two congruent triangles are related to each other by a sequence of isometries. It turns out that any two similar triangles are also related to each other by a sequence of isometries and a dilation. In other words, if two triangles are similar, then you can make one from the other using only isometries to move it around and a dilation to change the size.

Symmetry

Just what is symmetry? Symmetry is one of those things that you can recognize but cannot put into words. Many words or phrases have a meaning similar to symmetry; balanced and well-proportioned are two that immediately come to mind. However, those words don't help us when we are trying to come up with a mathematically precise definition of symmetry.

For example, take a look at the three triangles in Figure 25.7. The first triangle is a scalene triangle. The angles of the triangle are all distinct, as are the lengths of the sides. The second triangle is an isosceles triangle. It has more symmetry than the scalene triangle, because two of the sides and two of the angles are congruent. An isosceles triangle has more balance than the scalene triangle. The third triangle is an equilateral triangle. All three sides are congruent, as are all three angles. It is the most symmetrical triangle in the group, but why?

Figure 25.7

Three triangles with different levels of symmetry.

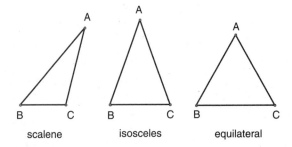

scalene   isosceles   equilateral

Imagine that you are small enough to stand at a vertex of these triangles. Suppose you start by standing on the vertices of the scalene triangle. Think about the views if you stand on vertex A, vertex B, and vertex C. Your view will change depending on which of the three vertices you are standing on. The segments that you look out on have different lengths. The included angles that the sides make have different measures in each case. There isn't much in common with the three perspectives, and that can be disconcerting. Each view involves something different.

What if you do the same thing with the isosceles triangle? Your view from vertex B and your view from vertex C are the same. The segment that you look out on has the same length in both views, and the measures of the angles formed by the included sides are the same. So there's some familiarity in what you see from vertex B when compared to what you see from vertex C. Things look the same from vertex B and vertex C, and the symmetry of this triangle is greater than the symmetry of the scalene triangle.

Moving on to the equilateral triangle, imagine your view from each of the three vertices. Because all sides are congruent and all angles are congruent, the view is the same from all three vertices. Identical views go hand in hand with maximum symmetry.

Now look at things from a different perspective. Instead of shrinking down to the triangle, imagine that you are all-powerful and are in complete control of the triangle. You can rotate it and flip it. If you rotate the scalene triangle, it looks different until you have gone full circle. Any way that you flip the scalene triangle it looks different. If you rotate the isosceles triangle, it also looks different until you have gone full circle, but if you flip it (interchanging vertices B and C) it looks like you haven't done anything. Whenever you can rotate or flip a figure and it looks like you haven't done anything, the figure you are playing with has some symmetry. The more ways you can move a figure and have it look like nothing has happened, the more symmetry the figure has. You are getting closer to how a mathematician describes symmetry.

A *symmetry* of an object is an isometry that moves the object back onto itself. In a symmetry, once the movement is complete it looks like nothing has been done to it. All objects have at least one symmetry; the identity isometry can always be applied.

But this is a boring symmetry, hardly worth mentioning. But because the identity isometry is an isometry, it must be mentioned. The symmetries of an object follow the isometries of an object, so the names of the symmetries will be similar to the names of the isometries. There is a translation symmetry, a reflection symmetry, a rotation symmetry, and a glide reflection symmetry.

Solid Facts

A **symmetry** of an object is an isometry that moves the object back onto itself.

Example 1: What are the symmetries of a square?

Solution: A square looks pretty symmetrical (from a nonmathematical perspective) so there are probably lots of symmetries around. Figure 25.8 will help sort out the symmetries of a square.

Figure 25.8

The symmetries of a square.

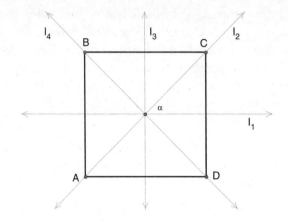

> ### Tangent Line
>
> There are some obvious and not-so-obvious applications of symmetry. Symmetry has obvious applications in architecture and art. The not-so-obvious applications include golf ball design and chemistry. The arrangement of the dimples of a golf ball affects a variety of properties of the ball, and the symmetry of a molecule affects its chemical properties.

First of all, look at reflections. If you reflect the square across l_1, you'll end up with what you started with. The same holds true for reflection across l_2, l_3, and l_4. So there are four reflection symmetries for a square.

Next, consider the rotations. If the center of the square is the point O (where the diagonals intersect), you can rotate the square 90°, 180°, 270°, and 360°. Realize that the last rotation is just the identity rotation.

So there are eight symmetries: four reflection symmetries and four rotation symmetries. That's a lot of symmetries, but you shouldn't be surprised, because the square is pretty symmetrical.

This is just the tip of the iceberg. I could write volumes on symmetry alone! The applications to art, music, construction, science, and nature are endless! And the mathematics behind symmetry are as elegant as the symmetry of a snowflake. I hope this brief introduction inspires you to learn more.

The Least You Need to Know

- The four types of isometries are translation, reflection, rotation, and glide reflection.

- A translation is a proper isometry, and translations other than the identity translation have no fixed points.

- A reflection is an improper isometry, and every point on the axis of reflection is a fixed point.

- A rotation is a proper isometry, and rotations have one fixed point.

- A glide reflection is an improper isometry, and glide reflections have no fixed points.

- The symmetry of a shape depends on the number of isometries that can be applied to the figure without changing how it looks. The more isometries there are, the more symmetrical the shape is.

Appendix A

Answer Key

This appendix provides the answers and solutions for the Put Me in, Coach! exercise boxes, organized by chapters.

Chapter 2

1. x = 3
2. x = 6

Chapter 4

1. x = 54, m∠Q = 162°, m∠P = 18°
2. ∠B ≅ ∠C

Chapter 6

1. 9 diagonals
2. 720°
3. 15
4. 120°
5. 12 sides

Chapter 7

1. The sequence: 6, 12, 20, 30, 42, ... From the first term to the second term you add 6. From the second to the third terms you add 8. From the third to the fourth terms you add 10. The nth term is given by the formula $(n + 1)(n + 2)$.

2. a. Statement; false

   b. Statement; true

   c. Not a statement

3. True

4. Truth table:

P	**Q**	$P \vee (\sim P \wedge Q)$
T	T	T
T	F	T
F	T	T
F	F	F

5. Truth table:

P	**Q**	$\sim P \rightarrow \sim Q$
T	T	T
T	F	T
F	T	F
F	F	T

6. Conditional: If the measure of $\angle A$ is less than 90°, then $\angle A$ is acute.

   Converse: If $\angle A$ is acute, then the measure of $\angle A$ is less than 90°.

   Inverse: If the measure of $\angle A$ is not less than 90°, then $\angle A$ is not acute.

   Contrapositive: If $\angle A$ is not acute, then the measure of $\angle A$ is not less than 90°.

Chapter 8

1. Yes

2. Theorem 8.3: If two angles are complementary to the same angle, then these two angles are congruent.

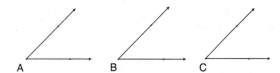

∠A *and* ∠B *are comple-mentary, and* ∠C *and* ∠B *are complementary.*

Given: ∠A and ∠B are complementary, and ∠C and ∠B are complementary.

Prove: ∠A ≅ ∠C .

Statements	Reasons
1. ∠A and ∠B are complementary, and ∠C and ∠B are complementary.	Given
2. $m\angle A + m\angle B = 90°$, $m\angle C + m\angle B = 90°$	Definition of complementary
3. $m\angle A = 90° - m\angle B$, $m\angle C = 90° - m\angle B$	Subtraction property of equality
4. $m\angle A = m\angle C$	Substitution (step 3)
5. ∠A ≅ ∠C	Definition of ≅

Chapter 9

1. If E is between D and F, then DE = DF – EF.

E *is between* D *and* F.

Given: E is between D and F

Prove: DE = DF – EF.

Statements	Reasons
1. E is between D and F	Given
2. D, E, and F are collinear points, and E is on $\overline{DF}$	Definition of between
3. DE + EF = DF	Segment Addition Postulate
4. DE = DF – EF	Subtraction property of equality

2. If $\overrightarrow{BD}$ divides $\angle ABC$ into two angles, $\angle ABD$ and $\angle DBC$, then
$m\angle ABD = m\angle ABC - m\angle DBC$.

$\overrightarrow{BD}$ *divides* $\angle ABC$ *into*
two angles, $\angle ABD$ *and*
$\angle DBC$.

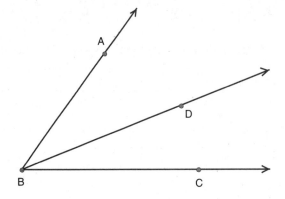

Given: $\overrightarrow{BD}$ divides $\angle ABC$ into two angles, $\angle ABD$ and $\angle DBC$.

Prove: $m\angle ABD = m\angle ABC - m\angle DBC$.

Statements	Reasons
1. $\overrightarrow{BD}$ divides $\angle ABC$ into two angles, $\angle ABD$ and $\angle DBC$	Given
2. $m\angle ABD + m\angle DBC = m\angle ABC$	Angle Addition Postulate
3. $m\angle ABD = m\angle ABC - m\angle DBC$	Subtraction property of equality

3. The angle bisector of an angle is unique.

$\angle ABC$ *with two angle*
bisectors: $\overrightarrow{BD}$ *and* $\overrightarrow{BE}$.

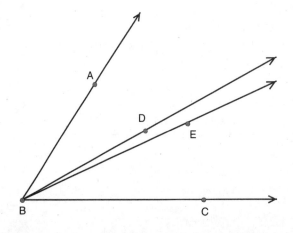

Given: ∠ABC with two angle bisectors: $\overrightarrow{BD}$ and $\overrightarrow{BE}$.

Prove: m∠DBE = 0 .

Statements	Reasons
1. $\overrightarrow{BD}$ and $\overrightarrow{BE}$ bisect ∠ABC	Given
2. ∠ABD ≅ ∠DBC and ∠ABE ≅ ∠EBC	Definition of angle bisector
3. m∠ABD = m∠DBC and m∠ABE = m∠EBC	Definition of ≅
4. m∠ABD + m∠DBE + m∠EBC = m∠ABC	Angle Addition Postulate
5. m∠ABD + m∠DBC = m∠ABC and m∠ABE + m∠EBC = m∠ABC	Angle Addition Postulate
6. 2m∠ABD = m∠ABC and 2m∠EBC = m∠ABC	Substitution (steps 3 and 5)
7. $m\angle ABD = \dfrac{m\angle ABC}{2}$ and $m\angle EBC = \dfrac{m\angle ABC}{2}$	Algebra
8. $\dfrac{m\angle ABC}{2} + m\angle DBE + \dfrac{m\angle ABC}{2} = m\angle ABC$	Substitution (steps 4 and 7)
9. m∠ABC + m∠DBE = m∠ABC	Algebra
10. m∠DBE = 0	Subtraction property of equality

4. The supplement of a right angle is a right angle.

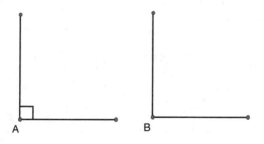

∠A and ∠B are supplementary angles, and ∠A is a right angle.

A   B

Given: ∠A and ∠B are supplementary angles, and ∠A is a right angle.

Prove: ∠B is a right angle.

Statements	Reasons
1. ∠A and ∠B are supplementary angles, and ∠A is a right angle	Given
2. $m\angle A + m\angle B = 180°$	Definition of supplementary angles
3. $m\angle A = 90°$	Definition of right angle
4. $90° + m\angle B = 180°$	Substitution (steps 2 and 3)
5. $m\angle B = 90°$	Algebra
6. ∠B is a right angle	Definition of right angle

Chapter 10

1. $m\angle 6 = 105°$, $m\angle 8 = 75°$

2. Theorem 10.3: If two parallel lines are cut by a transversal, then the alternate exterior angles are congruent.

l||m *cut by a transversal* t.

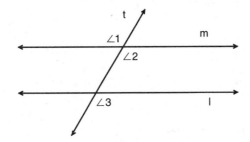

Given: l||m cut by a transversal t.

Prove: $\angle 1 \cong \angle 3$.

Statements	Reasons
1. l||m cut by a transversal t	Given
2. ∠1 and ∠2 are vertical angles	Definition of vertical angles
3. ∠2 and ∠3 are corresponding angles	Definition of corresponding angles
4. $\angle 2 \cong \angle 3$	Postulate 10.1
5. $\angle 1 \cong \angle 2$	Theorem 8.1
6. $\angle 1 \cong \angle 3$	Transitive property of ≅

3. Theorem 10.5: If two parallel lines are cut by a transversal, then the exterior angles on the same side of the transversal are supplementary angles.

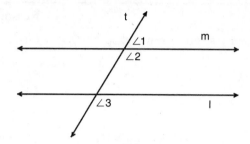

1 || m *cut by a transversal* t.

Given: 1 || m cut by a transversal t.

Prove: ∠1 and ∠3 are supplementary.

Statement	Reasons
1. 1 || m cut by a transversal t	Given
2. ∠1 and ∠2 are supplementary angles, and m∠1 + m∠2 = 180°	Definition of supplementary angles
3. ∠2 and ∠3 are corresponding angles	Definition of corresponding angles
4. ∠2 ≅ ∠3	Postulate 10.1
5. m∠2 = m∠3	Definition of ≅
6. m∠1 + m∠3 = 180°	Substitution (steps 2 and 5)
7. ∠1 and ∠3 are supplementary	Definition of supplementary

4. Theorem 10.9: If two lines are cut by a transversal so that the alternate exterior angles are congruent, then these lines are parallel.

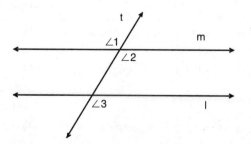

Lines 1 *and* m *are cut by a transversal* t.

Given: Lines l and m are cut by a transversal t, with $\angle 1 \cong \angle 3$.

Prove: l||m.

Statement	Reasons
1. Lines l and m are cut by a transversal t, with $\angle 1 \cong \angle 3$	Given
2. $\angle 1$ and $\angle 2$ are vertical angles	Definition of vertical angles
3. $\angle 1 \cong \angle 2$	Theorem 8.1
4. $\angle 2 \cong \angle 3$	Transitive property of $\cong$.
5. $\angle 2$ and $\angle 3$ are corresponding angles	Definition of corresponding angles
6. l||m	Theorem 10.7

5. Theorem 10.11: If two lines are cut by a transversal so that the exterior angles on the same side of the transversal are supplementary, then these lines are parallel.

Lines l and m are cut by a transversal t.

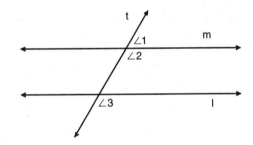

Given: Lines l and m are cut by a transversal t, $\angle 1$ and $\angle 3$ are supplementary angles.

Prove: l||m.

Statement	Reasons
1. Lines l and m are cut by a transversal t, and $\angle 1$ are $\angle 3$ supplementary angles	Given
2. $\angle 2$ and $\angle 1$ are supplementary angles	Definition of supplementary angles
3. $\angle 3 \cong \angle 2$	Example 2 in Chapter 4
4. $\angle 3$ and $\angle 2$ are corresponding angles	Definition of corresponding angles
5. l||m	Theorem 10.7

Chapter II

1. An isosceles obtuse triangle

2. The acute angles of a right triangle are complementary.

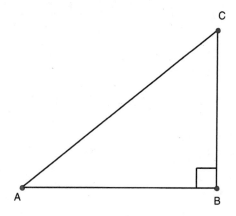

C   ΔABC *is a right triangle.*

Given: ΔABC is a right triangle, and ∠B is a right angle.

Prove: ∠A and ∠C are complementary angles.

Statement	Reasons
1. ΔABC is a right triangle, and ∠B is a right angle	Given
2. $m\angle B = 90°$	Definition of right angle
3. $m\angle A + m\angle B + m\angle C = 180°$	Theorem 11.1
4. $m\angle A + 90° + m\angle C = 180°$	Substitution (steps 2 and 3)
5. $m\angle A + m\angle C = 90°$	Algebra
6. ∠A and ∠C are complementary angles	Definition of complementary angles

3. Theorem 11.3: The measure of an exterior angle of a triangle equals the sum of the measures of the two nonadjacent interior angles.

$\triangle$ABC *with exterior angle*
$\angle$BCD.

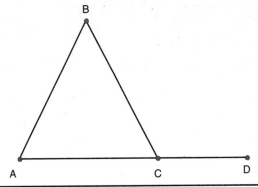

Statement	Reasons
1. $\triangle$ABC with exterior angle $\angle$BCD	Given
2. $\angle$DCA is a straight angle, and m$\angle$DCA $= 180°$	Definition of straight angle
3. m$\angle$BCA $+$ m$\angle$BCD $=$ m$\angle$DCA	Angle Addition Postulate
4. m$\angle$BCA $+$ m$\angle$BCD $= 180°$	Substitution (steps 2 and 3)
5. m$\angle$BAC $+$ m$\angle$ABC $+$ m$\angle$BCA $= 180°$	Theorem 11.1
6. m$\angle$BAC $+$ m$\angle$ABC $+$ m$\angle$BCA $=$ m$\angle$BCA $+$ m$\angle$BCD	Substitution (steps 4 and 5)
7. m$\angle$BAC $+$ m$\angle$ABC $=$ m$\angle$BCD	Subtraction property of equality

4. 12 units^2

5. 30 units^2

6. No, a triangle with these side lengths would violate the triangle inequality.

Chapter 12

1. Reflexive property: $\triangle$ABC $\cong$ $\triangle$ABC.

   Symmetric property: If $\triangle$ABC $\cong$ $\triangle$DEF, then $\triangle$DEF $\cong$ $\triangle$ABC.

   Transitive property: If $\triangle$ABC $\cong$ $\triangle$DEF and $\triangle$DEF $\cong$ $\triangle$RST, then $\triangle$ABC $\cong$ $\triangle$RST.

2. Proof: If $\overline{AC} \cong \overline{CD}$ and $\angle$ACB $\cong$ $\angle$DCB as shown in Figure 12.5, then $\triangle$ACB $\cong$ $\triangle$DCB.

Statement	Reasons
1. $\overline{AC} \cong \overline{CD}$ and $\angle ACB \cong \angle DCB$	Given
2. $\overline{BC} \cong \overline{BC}$	Reflexive property of $\cong$
3. $\triangle ACB \cong \triangle DCB$	SAS Postulate

3. If $\overline{CB} \perp \overline{AD}$ and $\angle ACB \cong \angle DCB$, as shown in Figure 12.8, then $\triangle ACB \cong \triangle DCB$.

Statement	Reasons
1. $\overline{CB} \perp \overline{AD}$ and $\angle ACB \cong \angle DCB$	Given
2. $\angle ABC$ and $\angle DBC$ are right angles	Definition of $\perp$
3. $m\angle ABC = 90°$ and $m\angle DBC = 90°$	Definition of right angles
4. $m\angle ABC = m\angle DBC$	Substitution (step 3)
5. $\angle ABC \cong \angle DBC$	Definition of $\cong$
6. $\overline{BC} \cong \overline{BC}$	Reflexive property of $\cong$
7. $\triangle ACB \cong \triangle DCB$	ASA Postulate

4. If $\overline{CB} \perp \overline{AD}$ and $\angle CAB \cong \angle CDB$, as shown in Figure 12.10, then $\triangle ACB \cong \triangle DCB$.

Statement	Reasons
1. $\overline{CB} \perp \overline{AD}$ and $\angle CAB \cong \angle CDB$	Given
2. $\angle ABC$ and $\angle DBC$ are right angles	Definition of $\perp$
3. $m\angle ABC = 90°$ and $m\angle DBC = 90°$	Definition of right angles
4. $m\angle ABC = m\angle DBC$	Substitution (step 3)
5. $\angle ABC \cong \angle DBC$	Definition of $\cong$
6. $\overline{BC} \cong \overline{BC}$	Reflexive property of $\cong$
7. $\triangle ACB \cong \triangle DCB$	AAS Theorem

5. If $\overline{CB} \perp \overline{AD}$ and $\overline{AC} \cong \overline{CD}$, as shown in Figure 12.12, then $\triangle ACB \cong \triangle DCB$.

Statement	Reasons
1. $\overline{CB} \perp \overline{AD}$ and $\overline{AC} \cong \overline{CD}$	Given
2. $\triangle ABC$ and $\triangle DBC$ are right triangles	Definition of right triangle
3. $\overline{BC} \cong \overline{BC}$	Reflexive property of $\cong$
4. $\triangle ACB \cong \triangle DCB$	HL Theorem for right triangles

6. If $\angle P \cong \angle R$ and M is the midpoint of $\overline{PR}$, as shown in Figure 12.17, then $\angle N \cong \angle Q$.

Statement	Reasons
1. $\angle P \cong \angle R$ and M is the midpoint of $\overline{PR}$	Given
2. $\overline{PM} \cong \overline{MR}$	Definition of midpoint
3. $\angle NMP$ and $\angle RMQ$ are vertical angles	Definition of vertical angles
4. $\angle NMP \cong \angle RMQ$	Theorem 8.1
5. $\triangle PMN \cong RMQ$	ASA Postulate
6. $\angle N \cong \angle Q$	CPOCTAC

Chapter 13

1. $x = 11$

2. $x = 12$

3. $40°$ and $140°$

4. If $\angle A \cong \angle D$ as shown in Figure 13.6, then $\dfrac{BC}{AB} = \dfrac{CE}{DE}$.

Statement	Reasons
1. $\angle A \cong \angle D$	Given
2. $\angle BCA$ and $\angle DCE$ are vertical angles	Definition of vertical angles
3. $\angle BCA \cong \angle DCE$	Theorem 8.1
4. $\triangle ACB \sim \triangle DCE$	AA Similarity Theorem
5. $\dfrac{BC}{AB} = \dfrac{CE}{DE}$	CSSTAP

5. 150 feet.

Chapter 14

1. If a line is parallel to one side of a triangle and passes through the midpoint of a second side, then it will pass through the midpoint of the third side.

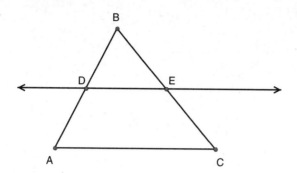

DE‖AC and D is the midpoint of AB.

Given: $\overline{DE}\|\overline{AC}$ and D is the midpoint of $\overline{AB}$.

Prove: E is the midpoint of $\overline{BC}$.

Statement	Reasons
1. $\overline{DE}\|\overline{AC}$ and D is the midpoint of $\overline{AB}$	Given
2. $\overline{DE}\|\overline{AC}$ and is cut by transversal $\overleftrightarrow{AB}$	Definition of transversal
3. $\angle BDE$ and $\angle BAC$ are corresponding angles	Definition of corresponding angles
4. $\angle BDE \cong \angle BAC$	Postulate 10.1
5. $\angle B \cong \angle B$	Reflexive property of $\cong$
6. $\triangle ABC \sim \triangle DBE$	AA Similarity Theorem
7. $\dfrac{DB}{AB} = \dfrac{BE}{BC}$	CSSTAP
8. $DB = \dfrac{AB}{2}$	Theorem 9.1
9. $\dfrac{DB}{AB} = \dfrac{1}{2}$	Algebra
10. $\dfrac{1}{2} = \dfrac{BE}{BC}$	Substitution (steps 7 and 9)
11. $BC = 2BE$	Algebra
12. $BE + EC = BC$	Segment Addition Postulate
13. $BE + EC = 2BE$	Substitution (steps 11 and 12)
14. $EC = BE$	Algebra
15. E is the midpoint of $\overline{BC}$	Definition of midpoint

2. $AC = 4\sqrt{3}$, $AB = 8\sqrt{3}$, $RS = 16$, $RT = 8\sqrt{3}$

3. $AC = 4\sqrt{2}$, $BC = 4\sqrt{2}$

Chapter 15

1. AD = 63, BC = 27, RS = 45

2. $\overline{AX}$, $\overline{CZ}$, and $\overline{DY}$

Trapezoid ABCD *with its four altitudes shown.*

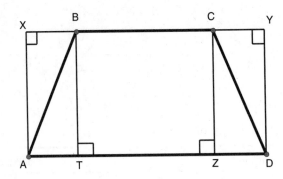

3. Theorem 15.5: In a kite, one pair of opposite angles is congruent.

Kite ABCD.

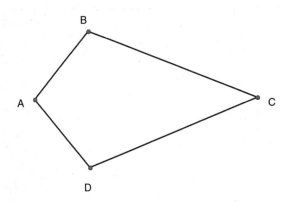

Given: Kite ABCD.

Prove: ∠B ≅ ∠D.

Statement	Reasons
1. ABCD is a kite	Given
2. $\overline{AB} \cong \overline{AD}$ and $\overline{BC} \cong \overline{DC}$	Definition of a kite
3. $\overline{AC} \cong \overline{AC}$	Reflexive property of ≅
4. △ABC ≅ △ADC	SSS Postulate
5. ∠B ≅ ∠D	CPOCTAC

4. Theorem 15.6: The diagonals of a kite are perpendicular, and the diagonal opposite the congruent angles bisects the other diagonal.

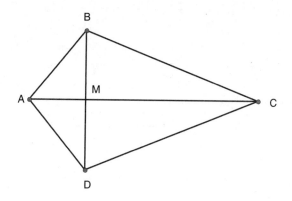

Kite ABCD.

Given: Kite ABCD.

Prove: $\overline{BD} \perp \overline{AC}$ and $\overline{BM} \cong \overline{MD}$.

Statement	Reasons
1. ABCD is a kite	Given
2. $\overline{AB} \cong \overline{AD}$ and $\overline{BC} \cong \overline{DC}$	Definition of a kite
3. $\overline{AC} \cong \overline{AC}$	Reflexive property of $\cong$
4. $\triangle ABC \cong \triangle ADC$	SSS Postulate
5. $\angle BAC \cong \angle DAC$	CPOCTAC
6. $\overline{AM} \cong \overline{AM}$	Reflexive property of $\cong$
7. $\triangle ABM \cong \triangle ADM$	SAS Postulate
8. $\overline{BM} \cong \overline{MD}$	CPOCTAC
9. $\angle BMA \cong \angle DMA$	CPOCTAC
10. $m\angle BMA = m\angle DMA$	Definition of $\cong$
11. $\angle BMD$ is a straight angle, and $m\angle BMD = 180°$	Definition of straight angle
12. $m\angle BMA + m\angle DMA = m\angle BMD$	Angle Addition Postulate
13. $m\angle BMA + m\angle DMA = 180°$	Substitution (steps 9 and 10)
14. $2m\angle BMA = 180°$	Substitution (steps 9 and 12)
15. $m\angle BMA = 90°$	Algebra
16. $\angle BMA$ is a right angle	Definition of right angle
17. $\overline{BD} \perp \overline{AC}$	Definition of $\perp$

5. Theorem 15.9: Opposite angles of a parallelogram are congruent.

Parallelogram ABCD.

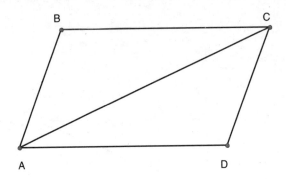

Given: Parallelogram ABCD.

Prove: $\angle ABC \cong \angle ADC$.

Statement	Reasons
1. Parallelogram ABCD has diagonal $\overline{AC}$.	Given
2. $\triangle ABC \cong \triangle CDA$	Theorem 15.7
3. $\angle ABC \cong \angle ADC$	CPOCTAC

6. 144 units²

7. 180 units²

8. Kite ABCD has area 48 units².

Parallelogram ABCD has area 150 units².

Rectangle ABCD has area 104 units².

Rhombus ABCD has area $\dfrac{35}{2}$ units².

Chapter 17

1. Circumference: 20π feet, length of $\overset{\frown}{RST} = \dfrac{155}{18}\pi$ feet

2. 9π feet²

3. 15π feet²

4. 28°

Chapter 18

1. Theorem 18.2: If two central angles are congruent, then the corresponding chords are congruent.

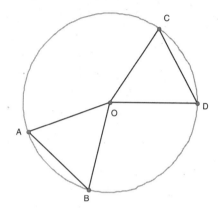

A circle with
$\angle AOB \cong \angle COD$.

Given: A circle with $\angle AOB \cong \angle COD$.

Prove: $\overline{AB} \cong \overline{CD}$.

Statement	Reasons
1. $\angle AOB \cong \angle COD$	Given
2. $\overline{OA} \cong \overline{OB} \cong \overline{OC} \cong \overline{OD}$	Theorem 17.1
3. $\triangle AOB \cong \triangle COD$	SAS Postulate
4. $\overline{AB} \cong \overline{CD}$	CPOCTAC

2. Theorem 18.4: If two arcs are congruent, then the corresponding chords are congruent.

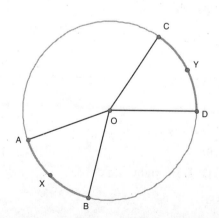

A circle with $\overset{\frown}{AXB} \cong \overset{\frown}{CYD}$.

Given: A circle with $\overset{\frown}{AXB} \cong \overset{\frown}{CYD}$.

Prove: $\overline{AB} \cong \overline{CD}$.

Statement	Reasons
1. $\overset{\frown}{AXB} \cong \overset{\frown}{CYD}$	Given
2. $\angle AOB \cong \angle COD$	Theorem 18.1
3. $\overline{OA} \cong \overline{OB} \cong \overline{OC} \cong \overline{OD}$	Theorem 17.1
4. $\triangle AOB \cong \triangle COD$	SAS Postulate
5. $\overline{AB} \cong \overline{CD}$	CPOCTAC

3. Theorem 18.6: If a radius is perpendicular to a chord, it bisects the arc of that chord.

A circle with $\overline{OC} \perp \overline{AB}$.

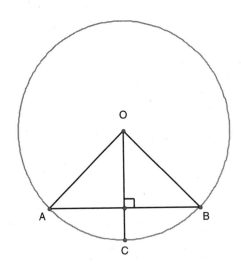

Given: A circle with $\overline{OC} \perp \overline{AB}$.

Prove: $\overset{\frown}{AXY} \cong \overset{\frown}{YZB}$.

Statement	Reasons
1. $\overline{OC} \perp \overline{AB}$	Given
2. $\overline{OA} \cong \overline{OB}$	Theorem 17.1
3. $\triangle AOB$ is an isosceles triangle	Definition of isosceles triangle
4. $\angle OAC \cong \angle OBC$	Theorem 12.1
5. $\angle OCA$ and $\angle OCB$ are right angles	Definition of $\perp$

Statement	Reasons
6. m∠OCA = 90° and m∠OCB = 90°	Definition of right angle
7. m∠OCA = m∠OCB	Substitution (step 6)
8. ∠OCA ≅ ∠OCB	Definition of ≅
9. ΔOAC ≅ ΔOBC	AAS Theorem
10. ∠AOC ≅ ∠BOC	CPOCTAC
11. m∠AOC = m∠BOC	Definition of ≅
12. m∠AOC = m$\widehat{AC}$ and m∠BOC = m$\widehat{BC}$	Central Angle Postulate
13. m$\widehat{AC}$ = m$\widehat{BC}$	Substitution (steps 11 and 12)
14. $\widehat{AC}$ ≅ $\widehat{CB}$	Definition of ≅

Chapter 19

1. Theorem 19.2: If two central angles are congruent, their intercepted arcs are congruent.

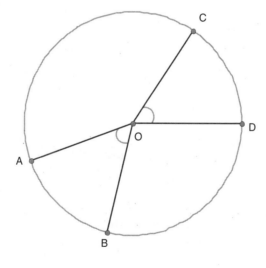

A circle with
∠AOB ≅ ∠COD .

Given: A circle with ∠AOB ≅ ∠COD .

Prove: $\widehat{AB}$ ≅ $\widehat{CD}$.

Statement	Reasons
1. $\angle AOB \cong \angle COD$	Given
2. $m\angle AOB = m\angle COD$	Definition of
3. $m\angle AOB = m\overset{\frown}{AB}$ and $m\angle COD = m\overset{\frown}{CD}$	Central Angle Postulate
4. $m\overset{\frown}{AB} = m\overset{\frown}{CD}$	Substitution (steps 2 and 3)
5. $\overset{\frown}{AB} \cong \overset{\frown}{CD}$	Definition of

2. $40°$

3. $69.5°$

4. $EC = 5$, $ED = 2$

5. $4\sqrt{7}$

Chapter 20

1. $\dfrac{3}{\sqrt{34}} = \dfrac{3\sqrt{34}}{34}$

2. $\dfrac{1}{\sqrt{3}} = \dfrac{\sqrt{3}}{3}$

3. tangent ratio $= \dfrac{\sqrt{40}}{3}$, sine ratio $= \dfrac{\sqrt{40}}{7}$

4. tangent ratio $= \dfrac{5}{\sqrt{56}} = \dfrac{5\sqrt{56}}{56}$, cosine ratio $= \dfrac{\sqrt{56}}{9}$

Chapter 23

1. $d = 3$

2. $d = \sqrt{18} = 3\sqrt{2}$

3. The midpoint is the point with coordinates (4.5, 6.5).

4. The slope is 1.

5. The equation of the line is $y = x + 2$.

6. The slope is 2, the y-intercept is 5.

7. The pairs of lines:

    a. neither

    b. perpendicular

    c. parallel

Postulates and Theorems

This appendix lists the postulates and theorems introduced in this book.

Postulates

Postulate 3.1: Every line contains at least two distinct points.

Postulate 3.2: Two points are contained in one and only one line.

Postulate 3.3: For all points A and B, AB = BA.

Postulate 3.4: The Ruler Postulate. The measure of any line segment is a unique positive number.

Postulate 3.5: The Segment Addition Postulate. If X is a point on $\overline{AB}$, then AX + XB = AB.

Postulate 3.6: If two points are in a plane, then the line containing these points is also in the plane.

Postulate 3.7: Every plane contains at least three noncollinear points.

Postulate 3.8: Three noncollinear points are contained in one and only one plane.

Postulate 4.1: The Protractor Postulate. The measure of an angle is a unique positive number.

Postulate 4.2: The Angle Addition Postulate. If a point D lies in the interior of ∠ABC, then $m\angle ABD + m\angle DBC = m\angle ABC$.

Postulate 5.1: The Parallel Postulate. Through a point not on a line, exactly one line is parallel to the given line.

Postulate 10.1: If two parallel lines are cut by a transversal, then the corresponding angles are congruent.

Postulate 12.1: SSS Postulate. If the three sides of one triangle are congruent to the three sides of a second triangle, then the triangles are congruent.

Postulate 12.2: SAS Postulate. If two sides and the included angle of one triangle are congruent to two sides and the included angle of a second triangle, then the triangles are congruent.

Postulate 12.3: The ASA Postulate. If two angles and the included side of one triangle are congruent to two angles and the included side of a second triangle, then the triangles are congruent.

Postulate 13.1: AAA Similarity Postulate. If the three angles of one triangle are congruent to the three angles of a second triangle, then the two triangles are similar.

Postulate 17.1: Central Angle Postulate. In a circle, the degree measure of a central angle is equal to the degree measure of its intercepted arc.

Postulate 17.2: Arc Addition Postulate. If B lies between A and C on a circle, then $m\widehat{AB} + m\widehat{BC} = m\widehat{AC}$.

Postulate 17.3: The ratio of the circumference of a circle to the length of its diameter is a unique positive constant.

Theorems

Theorem 8.1: When two lines intersect, the vertical angles formed are congruent.

Theorem 8.2: If two lines intersect to form consecutive congruent angles, then these lines are perpendicular.

Theorem 8.3: If two angles are complementary to the same angle, then these angles are congruent.

Theorem 9.1: The midpoint of a segment divides the segment into two pieces, each of which has length equal to one-half the length of the original segment.

Theorem 9.2: The midpoint of a segment is unique.

Theorem 9.3: The bisector of an angle divides the angle into two angles, each of which has measure equal to one-half the measure of the original angle.

Theorem 9.4: If ∠ABE is straight, ∠ABC, and ∠CBD is a right angle, then ∠ABC and ∠DBE are complementary. (Refer to Figure 9.6)

Theorem 9.5: The complement of an acute angle is an acute angle.

Theorem 9.6: The supplement of an acute angle is an obtuse angle.

Theorem 9.7: The bisectors of two adjacent supplementary angles form a right angle.

Theorem 10.1: Given a point A on a line l, there exists a unique line m perpendicular to l which passes through A.

Theorem 10.2: If two parallel lines are cut by a transversal, then the alternate interior angles are congruent.

Theorem 10.3: If two parallel lines are cut by a transversal, then the alternate exterior angles are congruent.

Theorem 10.4: If two parallel lines are cut by a transversal, then the interior angles on the same side of the transversal are supplementary angles.

Theorem 10.5: If two parallel lines are cut by a transversal, then the exterior angles on the same side of the transversal are supplementary angles.

Theorem 10.6: When two parallel lines are cut by a transversal, if one of the lines is perpendicular to the transversal, then both of the lines are perpendicular to the transversal.

Theorem 10.7: If two lines are cut by a transversal so that the corresponding angles are congruent, then these lines are parallel.

Theorem 10.8: If two lines are cut by a transversal so that the alternate interior angles are congruent, then these lines are parallel.

Theorem 10.9: If two lines are cut by a transversal so that the alternate exterior angles are congruent, then these lines are parallel.

Theorem 10.10: If two lines are cut by a transversal so that the interior angles on the same side of the transversal are supplementary, then these lines are parallel.

Theorem 10.11: If two lines are cut by a transversal so that the exterior angles on the same side of the transversal are supplementary, then these lines are parallel.

Theorem 11.1: In a triangle, the sum of the measures of the interior angles is 180°.

Theorem 11.2: Each angle of an equiangular triangle measures 60°.

Theorem 11.3: The measure of an exterior angle of a triangle equals the sum of the measures of the two nonadjacent interior angles.

Theorem 11.4: The Pythagorean Theorem. The square of the length of the hypotenuse of a right triangle is equal to the sum of the squares of the lengths of the legs.

Theorem 11.5: Triangle Inequality. The sum of the lengths of any two sides of a triangle is greater than the length of the third side.

Theorem 12.1: In an isosceles triangle, the angles opposite the congruent sides are congruent.

Theorem 12.2: The AAS Theorem. If two angles and a nonincluded side of one triangle are congruent to two angles and a nonincluded side of a second triangle, then the triangles are congruent.

Theorem 12.3: The HL Theorem for Right Triangles. If the hypotenuse and a leg of one right triangle are congruent to the hypotenuse and a leg of a second right triangle, then the triangles are congruent.

Theorem 13.1: AA Similarity Theorem. If two angles of one triangle are congruent to two angles of a second triangle, then the two triangles are similar.

Theorem 13.2: The SAS Similarity Theorem. If an angle of one triangle is congruent to an angle of a second triangle and the including sides are proportional, then the triangles are similar.

Theorem 13.3: The SSS Similarity Theorem. If three sides of one triangle are proportional to the three corresponding sides of another triangle, then the triangles are similar.

Theorem 14.1: The altitude drawn to the hypotenuse of a right triangle separates the right triangle into two right triangles that are similar to each other and to the original right triangle.

Theorem 14.2: If a line is parallel to one side of a triangle and intersects the other two sides, then it divides these sides proportionally.

Theorem 14.3: Every equiangular triangle is an equilateral triangle, and every equilateral triangle is an equiangular triangle.

Theorem 14.4: In a 30-60-90 triangle, the hypotenuse has a length equal to twice the length of the leg opposite the shorter angle, and the length of the leg opposite the 60° angle is the product of $\sqrt{3}$ and the length of the shorter leg.

Theorem 15.1: The median of a trapezoid is parallel to each base.

Theorem 15.2: The length of the median of a trapezoid equals one-half the sum of the lengths of the two bases.

Theorem 15.3: In a kite, one pair of opposite angles are congruent.

Theorem 15.4: The diagonals of a kite are perpendicular and one diagonal bisects the other.

Theorem 15.5: A diagonal of a parallelogram separates it into two congruent triangles.

Theorem 15.6: Opposite sides of a parallelogram are congruent.

Theorem 15.7: Opposite angles of a parallelogram are congruent.

Theorem 15.8: The diagonals of a parallelogram bisect each other.

Theorem 15.9: The diagonals of a rectangle are congruent.

Theorem 15.10: All sides of a rhombus are congruent.

Theorem 15.11: The diagonals of a rhombus are perpendicular.

Theorem 16.1: If one pair of opposite sides of a quadrilateral are parallel and congruent, then the quadrilateral is a parallelogram.

Theorem 16.2: If both pairs of opposite sides of a quadrilateral are congruent, then the quadrilateral is a parallelogram.

Theorem 16.3: If both pairs of opposite angles of a quadrilateral are congruent, then the quadrilateral is a parallelogram.

Theorem 16.4: If the diagonals of a quadrilateral bisect each other, then the quadrilateral is a parallelogram.

Theorem 16.5: If the diagonals of a parallelogram are congruent, then the parallelogram is a rectangle.

Theorem 16.6: If the diagonals of a parallelogram are perpendicular, the parallelogram is a rhombus.

Theorem 16.7: If the midpoints of the sides of a rectangle are joined in order, the quadrilateral formed is a rhombus.

Theorem 16.8: If the diagonals of a parallelogram are congruent and perpendicular, the parallelogram is a square.

Theorem 17.1: All radii of a circle are congruent.

Theorem 17.2: The measure of an inscribed angle of a circle is one-half the measure of its intercepted arc.

Theorem 17.3: An angle inscribed in a semicircle is a right angle.

Theorem 17.4: If two inscribed angles intercept the same arc, then these angles are congruent.

Theorem 18.1: If two chords of a circle are congruent, then the corresponding central angles are congruent.

Theorem 18.2: If two central angles are congruent, then the corresponding chords are congruent.

Theorem 18.3: If two chords of a circle are congruent, then the intercepted arcs are congruent.

Theorem 18.4: If two arcs are congruent, then the corresponding chords are congruent.

Theorem 18.5: If a radius is perpendicular to a chord, it bisects that chord.

Theorem 18.6: If a radius is perpendicular to a chord, it bisects the arc of that chord.

Theorem 18.7: The perpendicular bisector of a chord contains the center of the circle.

Theorem 19.1: If two arcs are congruent, their central angles are congruent.

Theorem 19.2: If two central angles are congruent, their intercepted arcs are congruent.

Theorem 19.3: The measure of an angle formed by two chords that intersect within a circle is found by taking the average of the degree measures of the arcs intercepted by the angle and its vertical angle.

Theorem 19.4: If two chords intersect within a circle, then the product of the lengths of the segments of one chord is equal to the product of the lengths of the segments of the other.

Theorem 19.5: If two chords of a circle are parallel, then the intercepted arcs between these chords are congruent.

Formulas

Several formulas were used in this book. I've summarized them here in one convenient, easy to find, location.

The total number of diagonals, D, in a polygon with n sides is given by the formula

$$D = \frac{n(n-3)}{2}.$$

The sum of the interior angles of a polygon having n sides is given by

$$S = (n-2) \times 180°.$$

Given an n-sided equiangular or regular polygon, the measure, I, of each interior angle is given by the formula

$$I = \frac{(n-2) \times 180°}{n}.$$

In a right triangle with legs having lengths a and b and the hypotenuse having length c, the Pythagorean Theorem states that

$$a^2 + b^2 = c^2.$$

The circumference, C, of a circle with radius r is given by the formula

$$C = 2\pi r.$$

On a circle, the arc length, l, of a sector having degree measure m can be found using the equation

$$l = \frac{m}{360°} C,$$

where C is the circumference of the circle.

The area, A, of a circle whose radius has length r is given by the equation

$$A = \pi r^2.$$

The area, S, of a sector with degree measure θ is found by evaluating

$$\frac{\theta \pi r^2}{360}.$$

The distance d between two points having coordinates (x,y) and (a,b) is given by the formula

$$d = \sqrt{(x-a)^2 + (y-b)^2}.$$

The midpoint of the line segment with endpoints having coordinates (a,b) and (c,d) is given by the formula

$$\left(\frac{a+c}{2}, \frac{b+d}{2} \right).$$

The slope of a line that passes through the points with coordinates (a,b) and (c,d) is given by the equation

$$\frac{d-b}{c-a}.$$

The Taxi-Cab distance between two points (a,b) and (c,d) is determined by the formula

$$d_{Taxi} = |a-c| + |b-d|.$$

The Max distance between two points (a,b) and (c,d) is determined by the formula

$$d_{Max} = Max\{|a-c|, |b-d|\}.$$

Glossary

acute angle   An angle whose measure is between 0° and 90°.

acute triangle   A triangle with all three angles having measure less than 90°.

addition property of equality   If x = y, then x + z = y + z.

addition property of inequality   If x > y, then x + z > y + z.

additive property of inequality   If a > b and c > d, then a + c > b + d.

adjacent angles   Two angles that share a common side and a common vertex, but have no interior points in common.

alternate exterior angles   Two exterior angles that have different vertices and lie on opposite sides of the transversal.

alternate interior angles   Two interior angles that have different vertices and lie on opposite sides of the transversal.

altitude of a trapezoid   A perpendicular line segment from a vertex of one base to the other base (or to an extension of that base).

altitude of a triangle   A line segment drawn perpendicularly from a vertex of the triangle to the opposite side of the triangle.

apex of a cone   The noncoplanar point that is joined to every point on the base.

apex of a pyramid   The noncoplanar point.

arc of a circle   The part of a circle determined by two points and all points on the circle between them.

area of a polygon   The measurement of the size of the region within the polygon. The units of area are the units of length squared.

base angles of a trapezoid   A pair of angles that share a common base.

base edges of a prism   The edges of the base of the prism.

base of a cone   The planar closed curve of the cone.

base of a prism   One of the congruent polygons that lie in the parallel planes.

base of a pyramid   The planar polygon whose vertices are all joined to the non-coplanar point.

base of a trapezoid   One of the parallel sides.

base of a triangle   The side opposite the vertex.

between   For any three collinear points A, B, and C, B is between A and C if AB + BC = AC.

bisector of an angle   A ray that divides the angle into two congruent angles.

center of a circle   The point equidistant from all points on the circle.

center of a sphere   The point that is a fixed distance from each point on the sphere.

chord of a circle   A line segment that joins two points of a circle.

circle   The set of all points in a plane that are a fixed distance from a given point.

circular cylinder   A cylinder with a circular base.

circumference of a circle   The linear measure of the distance around the circle.

collinear points   Points that lie on the same line.

complementary angles   Two angles whose sum is 90°.

concave polygon   A polygon in which at least one diagonal is outside the polygon.

concentric circles   Circles that have the same center.

cone   The solid formed by the interior of a planar closed curve together with the surface formed when each point on the closed curve is joined to a noncoplanar point using line segments.

congruent angles   Two angles with the same measure.

congruent arcs   Arcs on circles with congruent radii that have the same degree measure.

congruent circles   Circles that have congruent radii.

congruent segments   Two segments that have the same length.

congruent triangles   The six parts of one triangle are congruent to the six corresponding parts of the other triangle.

conjunction of P and Q   A statement of the form "P and Q."

consecutive angles of a polygon   Two angles that share a common side.

consecutive sides of a polygon   Two sides that share a common vertex.

consecutive vertices of a polygon   Two vertices that share a common side.

contrapositive of $P \rightarrow Q$ The statement $\sim Q \rightarrow \sim P$.

converse of $P \rightarrow Q$ The statement $Q \rightarrow P$.

convex polygon   A polygon in which all diagonals lie inside of the polygon.

coplanar points   Points that lie in the same plane.

corresponding angles   When two lines are cut by a transversal, they are the angles that lie in the same position relative to their respective line.

cosecant of an angle   The ratio of the length of the hypotenuse of a right triangle divided by the length of the side opposite the angle.

cosine of an angle   The ratio of the length of the side adjacent to the angle divided by the length of the hypotenuse of a right triangle.

cotangent of an angle   The ratio of the length of the adjacent side divided by the length of the side opposite the angle of a right triangle.

cylinder   The solid generated by two congruent closed curves in parallel planes together with the surface formed by line segments joining corresponding points of the two curves.

definition   A description that captures the essential qualities of the object you are trying to describe.

diagonal of a polygon   A line segment connecting any two nonconsecutive vertices of the polygon.

diameter of a circle   A chord that contains the center of the circle.

dilation   A transformation that changes the length of all line segments by the same proportion.

disjunction of P and Q   A compound statement of the form "P or Q".

division property of equality   If x = y and z ≠ 0, then $\frac{x}{z} = \frac{y}{z}$.

division property of inequality   If x > y and z > 0, then $\frac{x}{z} > \frac{y}{z}$.

endpoint of ray AB   The point A.

endpoints of a line segment   The two points where the line segment begins and ends.

equiangular polygon   A polygon with all angles congruent.

equiangular triangle   A triangle with all angles congruent.

equilateral polygon   A polygon with all sides congruent.

equilateral triangle   All three sides of the triangle are congruent.

equivalence relation   A relation that has reflexive, symmetric and transitive properties.

exterior angle of a triangle   The angle formed by a side and an extension of the adjacent side.

exterior angles   When two lines are cut by a transversal, they are the angles formed outside of the two lines.

fixed point of an isometry   A point that is its own image under an isometry.

glide reflection in the plane   An isometry that consists of a translation followed by a reflection.

greater than   If x and y are real numbers, x is greater than y if there exists a positive number z for which x = y + z. If x is greater than y, write x > y.

height of a triangle   The length of a triangle's altitude.

image of a point under an isometry   The final position of the point.

implication or **conditional statement**   A statement of the form "If P then Q."

inductive reasoning   The process of observing a pattern and drawing a general conclusion from those observations.

inscribed angle of a circle   An angle whose vertex is a point on the circle and whose sides are chords of the circle.

inscribed polygon of a circle   A polygon whose vertices are points on the circle and whose sides are chords of the circle.

interior angles   When two lines are cut by a transversal, they are the angles formed between the two lines.

invariant property of a transformation   A property that remains unchanged by a transformation.

inverse of $P \rightarrow Q$ The statement $\sim P \rightarrow \sim Q$.

isometry   A transformation that preserves the relative distance between points.

isosceles trapezoid   A trapezoid with congruent legs.

isosceles triangle   A triangle with at least two congruent sides.

kite   A quadrilateral with two distinct pairs of congruent adjacent sides.

lateral edges of a prism   The line segments connecting corresponding vertices of the bases of the prism.

lateral face of a prism   The quadrilateral formed by a pair of corresponding base edges and the lateral edges that connect the corresponding vertices.

legs of a trapezoid   The nonparallel sides.

length of a line segment   The distance between a line's endpoints.

line   A straight arrangement of points.

line segment AB   The set consisting of the two points A and B, and all the points between them that lie on the line containing A and B.

logically equivalent statements   Statements with the same truth values for all possible true/false combinations of their components.

major arc   An arc whose degree measure is between 180° and 360°.

median of a trapezoid   The line segment joining the midpoints of the two legs.

midpoint of a line segment   The point on the segment that divides the segment into two congruent segments.

minor arc   An arc whose degree measure is between 0° and 180°.

multiplication property of equality   If $x = y$, then $x \times z = y \times z$.

multiplication property of inequality   If $x > y$ and $z > 0$, then $x \times z > y \times z$.

noncollinear points   Points arranged so that there is no line that contains all of the points.

noncoplanar points   Points arranged so that there is no plane that contains all of the points.

oblique circular cone   A circular cone where the line segment connecting the apex of the cone to the center of the circular base is not perpendicular to the plane of the base.

oblique prism   A prism in which the angle formed between a lateral edge and a base edge is not a right angle.

obtuse angle   An angle whose measure is between 90° and 180°.

obtuse triangle   A triangle with one obtuse angle.

parallel lines   Lines that lie in the same plane and do not intersect.

parallelogram   A quadrilateral that has both pairs of opposite sides parallel.

perimeter of a triangle   The sum of the lengths of the three sides.

perpendicular lines   Two lines that intersect to form one right angle.

pi (π)   The ratio between the circumference of a circle and the length of a diameter of that circle.

point   The basic unit of geometry.

polygon   A closed plane figure whose sides are line segments that intersect only at the endpoints.

postulate   A basic statement that is assumed without proof to be self-evident.

prism   The region formed by two parallel congruent polygons having corresponding vertices joined by line segments.

proof   A sequence of statements, each supported by a reason, that starts with a given set of premises and leads to a valid conclusion.

pyramid   The region formed by joining the vertices of a polygon within a plane to a point outside of the plane.

radius of a circle   A line segment with one endpoint being the center of the circle, the other endpoint being a point on the circle.

radius of a sphere   The fixed distance between the center of a sphere and any point on the sphere.

ray AB   The part of line AB that contains the point A and all of the points on $\overleftrightarrow{AB}$ that are on the same side of point A as point B.

rectangle   A parallelogram that has a right angle.

reflection in the plane   An isometry that moves an object into a new position that is a mirror image of the original position.

reflexive property   x is related to x.

regular polygon   A polygon that is both equilateral and equiangular.

regular prism   A right prism whose base is a regular polygon.

relation   An operation that connects two elements of an associated collection of objects.

rhombus   A parallelogram with two congruent adjacent sides.

right angle   An angle whose measure is exactly 90°.

right circular cone   A circular cone where the line segment connecting the apex of the cone to the center of the circular base is perpendicular to the plane of the base.

right circular cylinder   A circular cylinder with its lateral surface perpendicular to the plane containing its base.

right prism   A prism in which the lateral edges are perpendicular to the base edges at the vertices.

right triangle   A triangle with one right angle.

rotation in the plane   An isometry that moves each point a fixed angle relative to a central point.

same-side exterior angles   Two exterior angles that lie on the same side of the transversal.

same-side interior angles   Two interior angles that lie on the same side of the transversal.

scalene triangle   A triangle in which none of the sides are congruent.

secant of an angle   The ratio of the length of the hypotenuse of a right triangle divided by the length of the adjacent side.

sector of a circle   The region bounded by two radii of the circle and the intercepted arc.

semicircle   An arc whose degree measure is exactly 180°.

side of a polygon   A line segment used to construct the polygon.

side of an angle   The two rays that meet to form the angle.

sine of an angle   The ratio of the length of the opposite side divided by the length of the hypotenuse of a right triangle.

space   The set of all points.

sphere   The set of all points (in three-dimensional space) that are a fixed distance r from a certain point.

square   A rectangle with congruent adjacent sides.

square root property of equality   Let x represent the length of a line segment, and let p represent a positive number. If $x^2 = p$, then $x = \sqrt{p}$.

statement   A group of words or symbols that can be classified collectively as true or false.

straight angle   An angle's measure is exactly 180°.

substitution property of equality   If a = b, then a replaces b in any equation.

subtraction property of equality   If x = y, then x – z = y – z.

subtraction property of inequality   If x > y, then x – z > y – z.

supplementary angles   Two angles whose sum is 180°.

symmetric property   If x is related to y, then y is related to x.

symmetry of an object   An isometry that moves the object back onto itself.

tangent of an angle   The ratio of the length of the opposite side of an angle divided by the length of the adjacent side of a right triangle.

tautology   A statement that is true for all possible truth values of its components.

theorem   A statement that can be proven using definitions, postulates, and other theorems.

transitive property   If x is related to y and y is related to z, then x is related to z.

translation in the plane   An isometry that moves every point in the plane a fixed distance in a fixed direction.

transversal line   A line that intersects two (or more) other lines at distinct points.

trapezoid   A quadrilateral with exactly two parallel sides.

truth table   A table that provides the truth values of a statement by considering all possible true/false combinations of the statement's components.

vertex of a polygon   A point of intersection of two sides of the polygon.

vertex of a prism   A point of intersection of two base edges.

vertex of an angle   The common endpoint of the two rays that form an angle.

vertical angles   A pair of nonadjacent angles formed by the intersection of two straight lines.

Index

Symbols

> (greater than symbol), 20
< (less than symbol), 20
|| (parallel symbol), 56
⊥ (perpendicular symbol), 54
π (Pi), 215
30-60-90 triangles, 174-176
45-45-90 triangles, 176-177
60-60-60 triangles, 173-174

A

AA Similarity Theorem, 162-163
AAA Similarity Postulate, 161
AAA Similarity Theorem, 163
AAS Postulate, 145-146
acute angles, 103-105
acute triangles, 127
adding
    angles, 41
    line segments, 32
additive property of equality, 17
additive property of inequality, 22-23
adjacent angles, 42-43
adjacent supplementary angle bisectors, 106
algebra
    circle exercises, 247-248
    properties of equality, 17-20
    properties of inequality, 20-23
    relations. *See* relations
algebraic facts, 8
alternate angles, 57, 113-114
altitudes
    similar triangles, 168-169
    trapezoids, 182
    triangles, 133
Angle Addition Postulate, 41
angles
    acute, 103-105
    adjacent, 42-43
    adjacent supplementary bisectors, 106
    alternate, 57, 113-114
    base, 180
    bisecting, 285-287
    bisector, 43, 101
    central, 213-214
    chords, 226-227, 240-241
    combining, 41
    complementary, 44
    congruent, 43
        circles, 238-241
        parallel line proofs, 119-120
        proving, 46-48, 149-150
    consecutive, 60
    corresponding, 57
    cosecants, 259-260
    cotangents, 259-260
    examples, 44-46
    exterior, 39, 57, 129-131
    inscribed, 211, 218-221
    interior, 39
        polygons, 64-68
        transversal lines, 57
        triangles, 128-129
    measuring, 40
    naming, 38-39
    nonacute, 265
    obtuse, 41
    parallelograms, 199
    Protractor Postulate, 40
    protractors, 283-284
    reflex, 38
    right, 41
    same-side exterior, 58
    same-side interior, 58
    secants, 259-260
    sides, 39
    straight, 41
    supplementary, 44, 115-116, 120-121
    transversal lines, 57
    triangles, 127
    vertexes, 39
    vertical, 51
apexes, 273-274
Archimedes, 5
arcs, 211-212, 215
    Central Angle Postulate, 214
    central angles, 213
    chords relationship, 227-228
    congruent, 212, 238-241
    intercepted, 213
    major, 212

measuring, 212
minor, 212
parallel, 246
areas
circles, 217
parallelograms, 190-191
rectangles, 191
rhombuses, 191
squares, 191
trapezoids, 189
triangles, 132-134
arguments
direct proofs, 87-89
indirect proofs, 89
law of detachment, 86-87
logical, 76
proofs. *See* proofs
ASA Postulate, 144-145
axioms, 8

B

base angles, 180
base edges, 271
bases
cones, 274
prisms, 271
pyramids, 273
trapezoids, 180
triangles, 133
bisecting
angles, 285-287
diagonals, 200-201
segments, 284-285
bisectors
adjacent supplementary
angles, 106
angles, 43, 101
perpendicular to chords,
231-233
Bolyai, Janos, 308

C

Cartesian coordinate system,
294
distance formula, 296-297
horizontal/vertical dis-
tances, 295-296
lines, graphing, 302
midpoint formula, 297-298
x-axis, 294
y-axis, 294
centers
circles, 210
spheres, 277
Central Angle Postulate, 214
central angles, 213-214
chords
angles, 226-227, 240-241
arcs relationship, 227-228
circles, 210
intersecting, 241, 244
lengths, 244
parallel, 245-246
radii relationship, 229-233
circles, 210
algebra exercises, 247-248
angles & chords relation-
ship, 226-227
arcs, 211-212, 215
Central Angle
Postulate, 214
central angles, 213
chords relationship,
227-228
congruent, 212
intercepted, 213
major, 212
measuring, 212
minor, 212
parallel, 246
area, 217
centers, 210

chords, 210
angles, measuring,
240-241
intersecting, 241, 244
lengths, 244
parallel, 245-246
circumference, 211,
215-217
compasses, 283
concentric, 211
congruent, 211
angles, 238-241
arcs, 238-241
inscribed angles theo-
rem, 221
constructing, 289
diameters, 210
examples, 233-236
inscribed angles, 211,
218-220
radii, 210, 229-233
sectors, 214
semicircles, 212, 220-221
shapes of, 313, 316
tangents, 217-218
three-dimensional,
277-278
trigonometric ratios,
263-264
unit, 313-315, 261-262
circular cones, 275
circular cylinders, 274
circumference, 211, 215-217
collinear points, 29
compasses, 283
complementary angles, 44,
102-105
compound statements
conditional, 79-81
conjunction, 77
disjunction, 78-79
concave polygons, 61
concentric circles, 211
conditional statements, 79-81

cones, 274-275
congruency
    angles, 43
        circles, 238-241
        parallel line proofs,
            119-120
        proving, 101-102,
            149-150
    arcs, 212, 238-241
    circles, 211
    diagonals
        rectangles, 201-202
        squares, 204-205
    line segments, 31
    segments, 149-150
    triangles, 140
        AAS Postulate, 145-146
        ASA Postulate, 144-145
        HL Theorem for Right
            Triangles, 146-148
        SAS Postulate, 142-144
        SSS Postulate, 141-142
conjunction statements, 77
consecutive angles, 60
consecutive vertices, 60
constructing
    circles, 289
    lines, 287-289
    quadrilaterals, 290-292
convex polygons, 61
coplanar points, 36
corresponding angles, 57
cosecants, 259-260
cosine ratios, 257-259
cotangents, 259-260
CPOCTAC (corresponding
    parts of congruent triangles
    are congruent), 140
    parallel lines, proving,
        150-152
    segment/angle congruence,
        proving, 149-150

cubes, 272, 278
cylinders, 274

D

deductive reasoning, 75-76
definitions
    proofs, writing, 90
    understanding, 10
    writing, 26
Descartes, Rene, 294
diagonals
    congruent, 201-202
    congruent & perpendicu-
        lar, 204-205
    parallelograms, 200-201
    perpendicular, 202
    polygons, 60-61
    spheres, 278
diameters (circles), 210
dilations, 325-326
direct proofs, 87-89
disjunction statements, 78-79
distances
    formula, 296-297
    max geometry, 312-313
    taxi-cab geometry, 310-312
division property of equality,
    18
division property of inequal-
    ity, 22
do-nothing translations, 320
dodecahedrons, 280

E

Elements of Euclid, 7
endpoints
    distance formula, 296-297
    horizontal/vertical segment
        lengths, 295-296
    line segments, 30

equality, properties of, 17
    additive, 17
    division, 18
    examples, 18-19
    multiplication, 18
    square root, 19-20
    substitution, 15
    subtraction, 17
equations (lines), 298-300
equiangular polygons, 67
equiangular triangles, 127
equilateral polygons, 67
equilateral triangles, 126
equivalence relations, 16, 21
Euclid, 7-9
Euclidean geometry, 305
Euler, Leonhard, 276
extended ratios, 156
exterior angles, 39, 57,
    129-131
extremes (triangles), 156

F

fixed points, 318
formal proofs, 92-95
formulas
    circles, 216-217
    distance, 296-297
    midpoint, 297-298
    point-slope, 300
    slope-intercept, 300
four-sided polygons. *See*
    quadrilaterals

G

Gauss, Carl, 5, 308
geometry
    real life applications, 4
given information (proofs), 90

glide reflections, 324-325
graphing lines, 302
greater than symbol (>), 20
Greek mathematicians
    Archimedes, 5
    Euclid, 7-9
    Thales, 6

H

heptadecogon, 5
Hilbert, David, 9
HL Theorem for Right
  Triangles, 146-148
horizontal distances (points),
  295-296
hyperbolic geometry, 307-309

I-J

icosahedrons, 279
identity reflections, 321
images (isometries), 318
implication statements. *See*
  conditional statements
inclusive or, 78
indirect proofs, 89
inductive reasoning, 74-75
inequality, properties of,
  20-21
    additive, 22-23
    division, 22
    equivalence relations, 21
    multiplication, 22
    subtraction, 22
inscribed angles
    circles, 211, 218-220
    congruent, 221
    semicircles, 220-221
inscribed polygons, 211
instincts (mathematical), 10

integers, 216
intercepted arcs, 213
interior angles, 39
    polygons, 64-68
    transversal lines, 57
    triangles, 128-129
intersecting lines, 50-51
invariants, 319
irrational numbers, 216
isometries, 318
    dilations, 325-326
    fixed points, 318
    glide reflections, 324-325
    images, 318
    invariants, 319
    reflections, 320-322
    rigid motion, 318
    rotations, 322-324
    translations, 319-320
isosceles triangles, 126

K-L

kites, 183, 290-291

lateral edges, 271
lateral faces, 271-273
law of detachment, 86-87
legs (trapezoids), 180
lengths
    chords, 244
    line segments, 30
    segments, 295-296
less than symbol (<), 20
lines, 50
    alternate angles, 113-114
    collinear points, 29
    constructing, 287
    defined, 28
    equations, 298-300
    graphing, 302

intersecting, 50-51
naming, 28
noncollinear points, 29
parallel, 55-56
    congruent angle proofs,
      119-120
    constructing, 289
    perpendicularity proofs,
      116-117
    proof by contradiction,
      118-119
    proofs, 112-113, 117
    slope, 301
    supplementary angle
      proofs, 115-116,
      120-121
    triangles, proving,
      150-152
perpendicular, 52-54
    constructing, 287-289
    parallelism proofs,
      116-117
    proofs, 110-112
    slope, 301
    uniqueness property,
      110-112
points, 28-29
segments, 29-30
    between points, 32-33
    bisecting, 284-285
    combining, 32
    congruent, 31
    distance formula,
      296-297
    endpoints, 30
    horizontal/vertical
      lengths, 295-296
    length, 30
    midpoints, 33
    naming, 30
    relations, 31-32
    unique midpoints, prov-
      ing, 99

tangents, 217-218
transversal, 57-58
Lobachevskian Postulate,
    308-309
Lobachevsky, Nikolai, 308
logical arguments, 76
logical equivalence, 81-82
logical tools (statements)
    conditional, 79-81
    conjunction, 77
    disjunction, 78-79
    negation, 76

M

major arcs, 212
mathematicians
    Bolyai, Janos, 308
    Euler, Leonhard, 276
    Gauss, Carl, 5, 308
    Greek
        Archimedes, 5
        Euclid, 7-9
        Thales, 6
    Hilbert, David, 9
    Lobachevsky, Nikolai, 308
    Riemann, Bernhard, 310
    Saccheri, Girolamo,
        307-309
    Wantzel, Pierre, 287
max geometry, 312-313
means (triangles), 156
Means-Extremes Property,
    156-159
measuring
    angles, 40
    arcs, 212
    chord angles, 240-241
    lines steepness, 298-299
    sizes, 5
medians (trapezoids), 180

midpoints
    formula, 297-298
    line segments, 33
    proving, 98-101
    rectangle sides, joining,
        203-204
    unique, 99
minor arcs, 212
motion, 322
multiplication property of
    equality, 18
multiplication property of
    inequality, 22

N

naming
    angles, 38-39
    lines, 28
    points, 27
    polygons, 64
    prisms, 272
    trapezoids, 180
natural numbers, 216
negating statements, 76
neutral geometry, 56
non-Euclidean geometry,
    305-307
nonacute angles, 265
noncollinear points, 29
noncoplanar points, 36
numbers, 216

O

objects
    dilations, 325-326
    glide reflections, 324-325
    reflections, 320-322
    rotations, 322-324
    sizes, 5

three-dimensional
    cones, 274-275
    cubes, 278
    cylinders, 274
    Platonic solids, 278-280
    polyhedrons, 276-277
    prisms, 270-272
    pyramids, 273
    rigid motion, 318
    spheres, 277-278
    translations, 319-320
oblique circular cones, 275
oblique prisms, 271
obtuse angles, 41
obtuse triangles, 127
octahedrons, 279

P

parallel lines, 55-56
    constructing, 289
    perpendicularity proofs,
        116-117
    proofs, 112-113, 117
        congruent angle proofs,
            119-120
        proof by contradiction,
            118-119
        supplementary angle
            proofs, 120-121
    slope, 301
    supplementary angle
        proofs, 115-116
    triangles, proving, 150-152
Parallel Postulate, 56, 307
parallel segments, 170-172
parallel symbol (||), 56
parallelograms, 184-186
    area, 190-191
    as quadrilaterals, 196
    congruent & perpendicular
        diagonals, 204-205

congruent diagonals,
201-202
constructing, 291
diagonals, bisecting,
200-201
midpoints of rectangle
sides, joining, 203-204
opposite sides congruent
and parallel, 196-197
perpendicular diagonals,
202
rectangles, 186-187
rhombuses, 187-189
squares, 189
two pairs of congruent
angles/sides, 197-199
patterns, 74-75
perimeters, 132
perpendicular diagonals
rhombuses, 202
squares, 204-205
perpendicular lines, 52-54
constructing, 287-289
parallelism proofs, 116-117
proofs, 110-112
slope, 301
uniqueness property,
110-112
perpendicular symbol (⊥), 54
Pi (π), 215
pictures, 11
planes, 34-35, 305
points, 35-36
rigid motion, 318
Platonic solids, 278-280
point-slope formula, 300
points
between, 32-33
Cartesian coordinate sys-
tem, 294
distance formula,
296-297

horizontal/vertical dis-
tances, 295-296
lines, graphing, 302
midpoint formula,
297-298
x-axis, 294
y-axis, 294
collinear, 29
coplanar, 36
endpoints
distance formula,
296-297
horizontal/vertical seg-
ment lengths, 295-296
line segments, 30
fixed, 318
lines, 28-29
midpoints
formula, 297-298
line segments, 33
proving, 98-101
rectangle sides, joining,
203-204
unique, proving, 99
naming, 27
noncollinear, 29
noncoplanar, 36
planes, 35
point-slope formula, 300
polygons
concave, 61
consecutive angles, 60
consecutive vertices, 60
convex, 61
diagonals, 60-61
equiangular, 67
equilateral, 67
four-sided. *See* quadrilaterals
inscribed, 211
interior angles, 64-66

naming, 64
regular, 67-69
rules, 62
sides, 60
three-dimensional,
276-277
three-sided. *See* triangles
vertices, 60
polyhedrons, 276-277
postulates. *See also* theorems
AAA Similarity, 161
AAS, 145-146
Angle Addition, 41
ASA, 144-145
Central Angle, 214
Lobachevskian, 308-309
parallel, 56, 307
proofs, writing, 90
protractor, 40
reading, 10
Riemannian, 310
SAS, 142-144
SAS Similarity, 164
SSS, 141-142
SSS Similarity, 164
prisms, 270-272
proofs
angles
complementary,
102-105
congruent, 101-102,
149-150
supplements, 105-107
circles
chord angles, measur-
ing, 240-241
congruent angles,
238-241
congruent arcs, 238-241
intersecting chords,
241, 244

parallel arcs, 246
parallel chords, 245-246
definitions, 90
direct, 87-89
formal, 92-95
given information, 90
indirect, 89
lines
  alternate angles,
    113-114
  parallel. *See* parallel
    lines, proofs
  perpendicular, 110-112
midpoints, 98-101
postulates, 90
proofs by contradiction,
  100
segment congruence,
  149-150
similar triangles, 161-164
theorems, 90
triangle congruency
  AAS Postulate, 145-146
  ASA Postulate, 144-145
  HL Theorem for Right
    Triangles, 146-148
  SAS Postulate, 142-144
  SSS Postulate, 141-142
two-column technique,
  90-91
unique midpoints, 99
writing, 6
properties
  equality, 17-20
  inequality, 20-23
  Means-Extremes, 156-159
  relations, 16-17
proportions
  segments, 170-172
  triangles, 156
Protractor Postulate, 40

protractors, 283-284
pyramids, 273
Pythagorean Theorem,
  134-135
  30-60-90 triangles, 175
  45-45-90 triangles, 176
  quadrilaterals, 191-193
  similar triangles, 169-170
  sine ratios, 255

Q

quadrilaterals, 63, 180
  constructing, 290
  diagonals, bisecting,
    200-201
  kites, 183, 290-291
  opposite sides congruent
    and parallel, 196-197
  parallelograms, 184-186,
    196
    area, 190-191
    constructing, 291
    rectangles, 186-187
    rhombuses, 187-189
    squares, 189
  Pythagorean Theorem,
    191-193
  rectangles, 291
  rhombuses, 291
  squares, 292
  trapezoids, 180-182
  two pairs of congruent
    angles/sides, 197-199

R

radii
  chords relationship,
    229-233
  circles, 210
  spheres, 277

rational numbers, 216
ratios
  cosecants, 259-260
  cosine, 257-259
  cotangents, 259-260
  extended, 156
  secants, 259-260
  sine, 255-257
  tangent, 252-255
  trigonometric, 263-265
rays, 33-34. *See also* angles
reading postulates, 10
real life applications, 4
reasoning
  deductive, 75-76
  inductive, 74-75
rectangles, 186-187
  area, 191
  congruent diagonals,
    201-202
  constructing, 291
  midpoints, joining,
    203-204
reflections, 320-322
  glide, 324-325
  identity, 321
reflex angles, 38
reflexive properties, 16
regular polygons, 67-69
regular prisms, 271
relations
  defined, 14
  equality, 15
  equivalence, 16, 21
  line segments, 31-32
  properties, 16-17
rhombuses, 187-189
  areas, 191
  constructing, 291
  midpoints of rectangle
    sides, joining, 203-204
  perpendicular diagonals,
    202

Riemann, Bernhard, 310
Riemannian Postulate, 310
right angles, 41
right circular cones, 275
right circular cylinders, 274
right prisms, 271
right triangles, 127
    cosine ratios, 258-259
    embedding in unit circles,
      261-262
    sine, 255-257
rigid motion, 318
rotations, 322-324
rows (truth tables), 88
rulers, 282
rules (polygons), 62

S

Saccheri, Girolamo, 9,
  307-309
same-side exterior angles, 58
same-side interior angles, 58
SAS Postulate, 142-144
SAS Similarity Postulate, 164
SAS Similarity Theorem, 164
scalene triangles, 126
secants, 259-260
sectors (circles), 214
segments
    bisecting, 284-285
    congruence, proving,
      149-150
    distance formula, 296-297
    horizontal/vertical lengths,
      295-296
    lines, 29-33
    midpoints
        formula, 297-298
        proving, 98-101
        unique, 99
    similar triangles, 170-172

semicircles, 212, 220-221
sides
    angles, 39
    parallelograms, 196-198
    polygons, 60
    triangles, 126
similar triangles, 159-161
    altitude theorem, 168-169
    proving, 161
        AA Similarity Theorem,
          162-163
        AAA Similarity
          Postulate, 161
        AAA Similarity
          Theorem, 163
        SAS Similarity
          Postulate, 164
        SAS Similarity
          Theorem, 164
        SSS Similarity
          Postulate, 164
        SSS Similarity
          Theorem, 164
        Pythagorean Theorem,
          169-170
    segments, 170-172
sine ratios, 255-257
size
    measuring, 5
    objects, 5
    triangles, 132-134
slopes
    lines, 298-299
    parallel lines, 301
    perpendicular lines, 301
    point-slope formula, 300
    slope-intercept formula,
      300
solids (Platonic), 278-280
space, 27
spheres, 277-278
spherical geometry, 309-310

square root property of equality, 19-20
squares, 189
    area, 191
    congruent & perpendicular
      diagonals, 204-205
    constructing, 292
    symmetry, 327-328
SSS Postulate, 141-142
SSS Similarity Postulate, 164
SSS Similarity Theorem, 164
statements, 75
    conditional, 79-81
    conjunction, 77
    disjunction, 78-79
    logically equivalent, 81-82
    negating, 76
    tautology, 82
steepness (lines), 298-299
straight angles, 41
straight edges, 282
substitution property of
  equality, 15
subtraction property of equality, 17
subtraction property of
  inequality, 22
supplementary angles, 44
    acute angles, 105
    parallel line proofs,
      115-116, 120-121
    proving, 105-107
symmetry, 326-328
    properties, 16
    squares, 327-328
synonyms, 26

T

tables (truth), 76-77
    conditional statements, 80
    conjunction statements, 77

direct proofs, 88
disjunction statements, 78-79
indirect proofs, 89
law of detachment, 86-87
negation, 77
rows, 88
tangent ratios, 252-255
tangents, 217-218
tautology, 82
taxi-cab geometry, 310-312
tetrahedrons, 278
Thales, 6
theorems. *See also* postulates
  AA Similarity, 162-163
  AAA Similarity, 163
  defined, 8
  HL Theorem for Right Triangles, 146
  proofs, writing, 90
  Pythagorean, 134-135
    30-60-90 triangles, 175
    45-45-90 triangles, 176
    quadrilaterals, 191-193
    sine ratios, 255
  SAS Similarity, 164
  similar triangles, 168-170
  SSS Similarity, 164
  Triangle Inequality, 135-137
  understanding, 11
three-dimensional objects
  cones, 274-275
  cubes, 278
  cylinders, 274
  Platonic solids, 278-280
  polyhedrons, 276-277
  prisms, 270-272
  pyramids, 273
  rigid motion, 318
  spheres, 277-278
three-sided polygons. *See* triangles

tools, 282-284
transformations (dilations), 325-326
transitive properties, 16
translations, 319-320, 324-325
transversal lines, 57-58
trapezoids, 180-182
  altitudes, 182
  area, 189
  bases, 180
  legs, 180
  medians, 180
  naming, 180
Triangle Inequality Theorem, 135-137
triangles, 63, 126
  30-60-90, 174-176
  45-45-90, 176-177
  60-60-60, 173-174
  acute, 127
  altitudes, 133
  angles, 127, 149-150
  area, 132-134
  base, 133
  congruence, 140
    AAS Postulate, 145-146
    ASA Postulate, 144-145
    HL Theorem for Right Triangles, 146-148
    SAS Postulate, 142-144
    SSS Postulate, 141-142
  CPOCTAC, 140
  equiangular, 127
  equilateral, 126
  extended ratios, 156
  exterior angles, 129-131
  extremes, 156
  height, 133
  interior angles sum, 128-129
  isosceles, 126

means, 156
Means-Extremes Property, 156-159
obtuse, 127
parallel lines, proving, 150-152
perimeter, 132
proportions, 156
Pythagorean Theorem, 134-135
right, 127
  cosine ratios, 258-259
  embedding in unit circles, 261-262
  sine, 255-257
scalene, 126
segment congruence, 149-150
sides, 126
similar, 159-161
  altitude theorem, 168-169
  Pythagorean Theorem, 169-170
  segments, 170-172
similarity proofs, 161-164
size, 132
Triangle Inequality Theorem, 135-137
triangular prisms, 271
trigonometry
  circles, 263-264
  cosecants, 259-260
  cosine ratios, 257-259
  cotangents, 259-260
  nonacute angles, 265
  secants, 259-260
  sine ratios, 255-257
  tangent ratios, 252-255
truth tables, 76-77
  conditional statements, 80
  conjunction statements, 77
  direct proofs, 88

disjunction statements,
  78-79
indirect proofs, 89
law of detachment, 86-87
negation, 77
rows, 88
two-column technique
 (proofs), 90-91

U–V

unit circles
    Euclidean geometry,
      313-314
    Max geometry, 315
    right triangles, embedding,
      261-262
    taxi-cab geometry, 314

vertexes, 39
vertical angles, 51
vertical distances (points),
 295-296
vertices, 60, 271

W–X–Y–Z

Wantzel, Pierre, 287
writing
   definitions, 26
   proofs, 6
       definitions, 90
       formal, 92-95
       given information, 90
       postulates, 90
       theorems, 90
       two-column technique,
         90-91

x-axis, 294

y-axis, 294